T0328420

Pathophysiology of Ischemia Reperfusion Injury and Use of Fingolimod in Cardioprotection

Pathophysiology of Ischemia Reperfusion Injury and Use of Fingolimod in Cardioprotection

NASEER AHMED

Assistant Professor, Department of Biological and Biomedical Sciences, Aga Khan University Medical College, Karachi, Pakistan

ACADEMIC PRESS

An imprint of Elsevier

Academic Press is an imprint of Elsevier
125 London Wall, London EC2Y 5AS, United Kingdom
525 B Street, Suite 1650, San Diego, CA 92101, United States
50 Hampshire Street, 5th Floor, Cambridge, MA 02139, United States
The Boulevard, Langford Lane, Kidlington, Oxford OX5 1GB, United Kingdom

British Library Cataloguing-in-Publication Data
A catalogue record for this book is available from the British Library

Library of Congress Cataloging-in-Publication Data
A catalog record for this book is available from the Library of Congress

ISBN: 978-0-12-818023-5

For Information on all Academic Press publications
visit our website at https://www.elsevier.com/books-and-journals

Publisher: Stacy Masucci
Acquisition Editor: Katie Chan
Editorial Project Manager: Sandra Harron
Production Project Manager: Sreejith Viswanathan
Cover Designer: Mark Rogers

Typeset by MPS Limited, Chennai, India

Working together
to grow libraries in
developing countries

www.elsevier.com • www.bookaid.org

Dedication

Special dedication to my beloved Father Hakeem Pir Muhammad (1944−2009) and mother for their continuous support and prayers.

Dedication

Contents

About the author

Naseer Ahmed is a clinician and scientist with training in both clinical and basic Cardiology and Clinical Pharmacology. He received his medical degree from the Islamic International Medical College, Riphah International University, Islamabad, Pakistan. After graduation, his research passion led him to the University of Verona, Italy where he excelled in basic, clinical and translational research and was awarded the title "Europeaus Doctorate" by European Union in addition to his PhD cardiovascu-
lar sciences and was recognized as the "Best Graduate Student" by the School of Life and Health Sciences. He has also worked as a postdoctoral fellow Clinical Pharmacology at the University of Verona, Verona, Italy. He was awarded the University of Verona PhD grant, Coperint International Mobility grant, ESC Educational grant, and Higher Education Commission Pakistan travel and research grants to initiate cutting-edge research in Pakistan. He has published more than 25 articles and abstracts in well-reputed international journals. He has active research collaboration and links with Italian, German, American, and British universities/research institutes. He served as a member of the advisory board for the PhD program at the School of Life and Health Sciences at the University of Verona. He is an active member of different professional associations including the Pakistan Cardiac Society, European Society of Cardiology and the American College of Clinical Pharmacology. Recently, he was the principal investigator in three clinical trials studying an innovative synthetic compound for cardioprotection. He is currently working at the Aga Khan University-Karachi as an assistant professor in the Department of Biological and Biomedical Sciences where he contributes to undergraduate and postgraduate teaching and research supervision.

About the author

Foreword

This book addresses the issue of ischemia-reperfusion injury and provides insights on new mechanisms and potential therapeutic approaches. The idea to write a book on this neglected but important issue came from the direct scientific experiences of distinguished Pakistani scientist Dr. Naseer Ahmed. This resource is of value to both basic researchers, including PhD students and clinicians, and to practitioners as well as to pharmacologists, cardiologists, and cardiac surgeons.

This book is divided into four parts: the first part addresses ischemia as a general process, the second part focuses on the cardiac events of ischemia-reperfusion damage, the third discusses the therapeutic aspects and problems of the injury, and the last part addresses the role of fingolimod in cardioprotection.

The first part of the book is an extensive and accurate review of the processes leading to ischemia-reperfusion injury; the literature cited in Chapter 1 is abundant, and with the help of clear schematic drawings, can be of great help to graduate students and researchers addressing for the first time this complex phenomenon. With a broad perspective, this part focuses on events at the cellular and tissue levels, but is not limited to cardiac tissue.

In the second part, the authors go into more detail about cardiologic problems. In Chapter 2 cardiac ischemia is reviewed followed by a discussion of the reperfusion-associated events occurring in different conditions, including surgery. Analysis of the molecular events leading to cell damage and some methods to limit their extensions is also included in Chapter 3.

The last part, which includes Chapters 4 and 5, is dedicated to the therapeutic aspects of the problem. Chapter 4 deals with cardioprotection at large, with analysis of the approaches to dealing with oxidative stress and apoptosis and a brief summary of the pharmacological approaches to limiting specific mechanisms of myocardial cell damage. The last chapter is the most interesting as it focuses on novel approaches for controlling postischemic cell damage with special emphasis on modulation of the sphingosine system. In this chapter, the pharmacological modulation of the system and the use of fingolimod are described in detail and to provide useful information to both researchers and clinicians. The basic mechanisms of the cardiovascular effects of fingolimod are described along with clinical perspectives.

The multidisciplinary approach used to address the different aspects of this complex medical issue makes *Pathophysiology of Ischemia Reperfusion Injury and Use of Fingolimod in Cardioprotection* a relevant book for a wide range of professionals working in cardiology, from basic researchers interested in general phenomena to clinical researchers and finally to practitioners attracted by the applicative aspects underlined in this book.

Guido F. Fumagalli

Department of Diagnostics & Public Health, School of Medicine,
University of Verona, Verona, Italy

Verona, August 13, 2018

Acknowledgments

First and foremost, thanks to *Allah Almighty*, for the guidance and help in giving me the strength to complete this book.

I would like to express my deep and sincere gratitude to my supervisors Prof. Giuseppe Faggian, Prof. Enrico Barbieri, Prof. Giovanni B. Luciani, and Dr. Alessio Rungatscher whose responsible mentorship and constant guidance helped me develop my scientific thinking. I would like to show appreciation to my colleague, Dr. Soban Sadiq, for his support and guidance.

Thanks also to my beloved mother, my dear brother Dr. Muhammad Nazeer and his wife Dr. Nargis, my sisters Kishwer, Nusrat, Samina, and Saima, and my sweet younger sister Engr. Asima who have always prayed, supported, encouraged, and believed in me in all my endeavors. In addition, I cannot forget the continuous support of my friends and colleagues Dr. Giulio, Dr. Guido, Dr. Alessandro, and Dr. Arfa. I would like to thank my lovely wife, Dr. Adeela Mehmood, for making my dream, of publishing book, a reality.

I must appreciate Prof. Muhammad Perwaiz Iqbal and Dr. Rehana Rehman for encouraging me and providing an excellent environment to work on this book and Prof. Giudo Fumagalli and Anwar-ul-Hassan Gilani for their mentorship in the field of clinical pharmacology.

Naseer Ahmed

CHAPTER 1

Introduction to ischemia−reperfusion injury

Abstract

The extent of cell dysfunction, injury, and/or death is dependent on both the intensity and the duration of ischemia. In this aspect, revascularization and restoration of blood flow as soon as possible remains the mainstay of all current therapeutic strategies for ischemia. However, not all organs demonstrate equal susceptibility to ischemia. Moreover, it now seems clear that reperfusion, although necessary to reestablish delivery of oxygen and nutrients to support cell metabolism and remove potentially damaging by-products of cellular metabolism, can elicit pathogenetic processes that exacerbate injury due to ischemia per se and may produce tissue injury in distant organs as a result of mediators released into the bloodstream, draining to vascularized tissues and subsequent delivery to remote organs. There are multiple events involving ischemia reperfusion−related injury.

Keywords: Ischemia; reperfusion injury; revascularization; mediators

Contents

Pathophysiology of Ischemia Reperfusion Injury and Use of Fingolimod in Cardioprotection
DOI: https://doi.org/10.1016/B978-0-12-818023-5.00001-7

Overview of ischemia/reperfusion

The amount of cell dysfunction, damage, and/or death relates to both the intensity and the duration of ischemia. Immediate revascularization and reestablishment of blood flow is still the most important part of all present therapies for ischemia. However, not all organs exhibit the same vulnerability to ischemia. It is evident that reperfusion, is although necessary to reestablish delivery of oxygen and nutrients to support cell metabolism and remove possible harmful by-products of cellular metabolism, can lead to pathogenetic processes that aggravate the damage associated with ischemia. It might damage tissue in distant organs due to the release of mediators into the blood flowing from vascularized tissues and later delivered to remote organs. Also, short bouts of ischemia/reperfusion (I/R) (ischemic preconditioning) before lethal ischemia start cell-survival programs that restrict postischemic damage. This discovery highlights that the reaction to ischemia is bimodal.

Due to ischemia, hypoxia develops which shifts the cell to anaerobic mode. This forms lactate and reduces the pH of the cell due to the collection of hydrogen ions. In compensation, the cell leads to outflow of hydrogen ions in exchange for sodium influx via the Na/H exchanger as a result;

- There is a reduction of cellular adenosine tri-phosphate (ATP) resulting in inactivation of ATPase, decreasing calcium efflux and restricting ER calcium reuptake causing calcium overload.
- There is more reduction in ATP formation because of the opening of the mitochondrial transition pore that damages the mitochondrial membrane potential.
- In the heart, intracellular proteases are activated along with these cellular changes. This harms the myofibrils and forms hypercontractures and contracture-band necrosis.

These cellular alterations, and as a result the amount of tissue damage, rely on the magnitude of ischemia and duration of the ischemic period.

Other biochemical events happen during ischemia that do not lead to ischemic injury per se, but when the blood flow is reestablished, they are energized by the supply of oxygen and produced elements in the blood, and can trigger a cascade of events that aggravate tissue injury (Fig. 1.1).

Although immediate reperfusion returns the supply of oxygen and substrates needed for aerobic formation of ATP and returns the extracellular pH to normal by washing out increased $H+$, reperfusion seems to

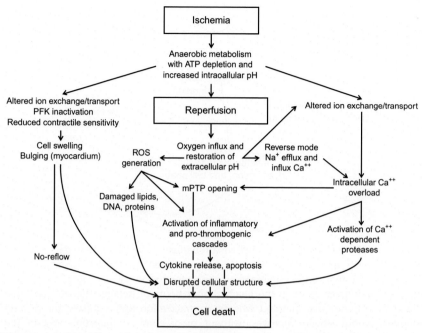

Figure 1.1 Cascades of ischemia reperfusion injury, Adapted from Theodore Kalogeris et al. (2012).

have harmful results (Figs. 1.1 and 1.2). This concept initially emerged more than 50 years ago, when an observation was made that reperfusion speeded up the formation of necrosis in hearts exposed to coronary ligation (Jennings, Sommers, Smyth, Flack, & Linn, 1960). This is called reperfusion injury to explain the processed related to restoring the blood supply. Reperfusion injury had not happened during the preceding ischemic period and can be stopped by an intervention administered only at the time of reperfusion. The presence of such lethal reperfusion injury as being different from the injury resulting previously from ischemia is still debated. However, interventions during myocardial reperfusion can truly decrease infarct size by up to 50%, supporting reperfusion phase-specific harmful events (Yellon & Hausenloy, 2007). The processes associated with reperfusion injury are complicated, multifactorial, and include (1) formation of ROS because of molecular oxygen being introduced again when the blood supply is reestablished, (2) calcium overload, (3) opening of the mitochondrial permeability transition (MPT) pore, (4) endothelial

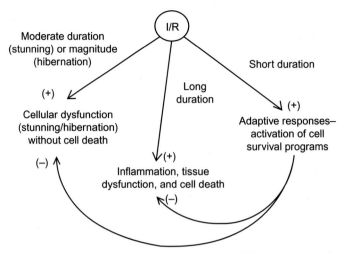

Figure 1.2 Tissue reactions to ischemia/reperfusion (I/R) are bimodal, and are affected by the time period of ischemia. Long duration and intense ischemia results in cell damage that can cause infarction, with reperfusion usually aggravating tissue damage via inflammation. In the heart, shorter bouts of ischemia (5–20 min duration) induce myocardial stunning, wherein contractile function is initially impaired on reperfusion, but slowly improves, without progression to infarction and in the absence of significant inflammation. whereas being subjected for a prolonged exposure to subacute levels of ischemia in the absence of reperfusion might result in myocardial hibernation with cardiac cells returning to a more primitive metabolic phenotype to survive but with decreased mechanical function. On the other hand, short periods of ischemia (<5 min) and then reperfusion (ischemic conditioning) stimulate cell-survival programs that restrict the magnitude of damage caused by being subjected to I/R for a long time period.

dysfunction, (5) appearance of a prothrombogenic phenotype, and (6) significant inflammation (Yellon & Hausenloy, 2007; Fig. 1.1).

It is obvious then that total damage to a tissue is the sum of impairment caused by ischemia and reperfusion (Figs. 1.1 and 1.2). It is also clear that the reperfusion phase is very dynamic and cell death can persist for up to 3 days after the start of reperfusion (Zhao et al., 2000). Therefore understanding the mechanisms involved leads to the establishment of appropriate therapeutic methods that decrease the amount of damage caused by I/R and increase the a myocardial tissue preservation (Fig. 1.2). This point is important for organ transplantation, cardiopulmonary bypass, and operation in a bloodless field.

Bimodal tissue response to ischemia/reperfusion

All tissues can bear varying short periods of ischemia in the absence of detectable functional deficits or evident damage (Fig. 1.2). Once a critical duration of ischemia has passed, which relies on the cell type and organ, cell injury and/or death results. Therefore we can conclude that the responses to ischemia are harmful, with reperfusion exacerbating the amount of tissue injury. However, Murry, Jennings, and Reimer Reimer (1986) found that subjecting the heart (or other tissues) to short bouts of ischemia and reperfusion (ischemic preconditioning) before a long duration of decrease in coronary blood flow (index ischemia) leads to powerful infarct-sparing results. This is an interesting discovery relating to therapeutic strategies that might prove effective in reducing the risk for and/or outcome of harmful cardiovascular events. Also, the finding of ischemic preconditioning demonstrates that the reaction to ischemia is bimodal: longer duration of ischemia results in cell dysfunction and/or death that is aggravated by reperfusion, while short cycles of conditioning ischemia have a protective role, making tissues resistant to the deleterious effects of prolonged ischemia and later reperfusion occurs by activating intrinsic cell-survival programs (Fig. 1.3).

Ischemia/reperfusion-induced stunning and hibernation versus irreversible cell damage and death

Contractile abnormalities in postischemic myocardial tissue were at one point considered to be a consequence of permanent cell injury and loss of normal myocardium. Currently, it is clear that despite reestablishment of normal coronary blood supply, mechanical damage can persist after reperfusion in the absence of permanent damage. For instance, in myocardial stunning short postischemic contractile dysfunction leads to lack of irreversible injury. It is not a result of a primary deficit in reperfusion (i.e., postischemic flow is normal or near normal; Bolli & Marban, 1999; Depre & Vatner, 2005; Fig. 1.2). Myocardial stunning apparently happens due to reperfusion. This stimulates the production of ROS (oxygen paradox); transient calcium overload related to low responsiveness of contractile elements to calcium (calcium paradox); activation of calpains, which catalytically cleave myofibrils; and changed membrane ion channel performance after immediate restoration of extracellular pH (pH paradox). Some medical investigators have stated that the stunning-induced deficits in contractile activity might provide protection. It restricts the progression

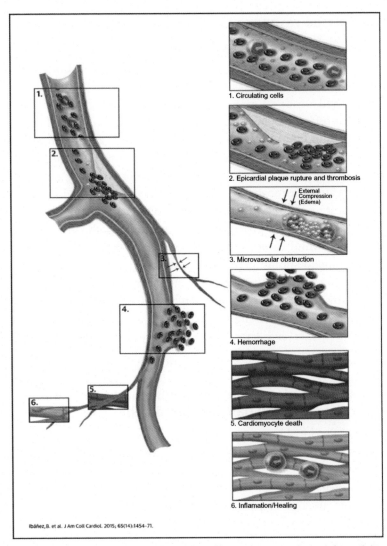

Figure 1.3 Players involved in ischemia/reperfusion injury. *Adapted from Ibáñez, B., Heusch, G., Ovize, M., & Van de Werf, F. (2015). Evolving therapies for myocardial ischemia/reperfusion injury. Journal of the American College of Cardiology, 65(14), 1454–1471.*

of the harsh cellular milieu caused by ischemia toward permanent injury during reperfusion, thus increasing the chance of cell survival (Bolli & Marban, 1999; Depre & Vatner, 2005). Myocytes subjected to long periods or repetitive intermittent ischemia may lead to another adaptive reaction by restoring to neonatal metabolic phenotype, supporting the utilization of carbohydrates as a fuel source. This phenomenon, wherein ischemic myocytes undergo a metabolic change to a glycolytic phenotype with decreased contractile function and energy requirements, is called myocardial hibernation (Depre & Vatner, 2005; Slezak et al., 2009). Such hibernation permits myocardial cells to resist decrease in oxygen and nutrient supply concomitant with subacute levels of ischemia. Since contractile function is limited, irreversible cardiomyocyte damage is absent. Reprogramming of cell metabolism, resulting in a reduction in energy usage and increased expression of stress and angiogenic proteins, causes the process associated with the assumption of ancestral phenotype. Typical cell remodeling alterations also happen in hibernating myocardium, having the look of polymorphic mitochondria, with a high level of lysosomes and reduced myofibrils. High vacuolar density and debris are consistent with a programmed cell death that results in a high survival of hibernating viable cells in ischemic organs by taking out nonfunctional cells. Hibernating myocardium also consists of apoptotic cells. Although these adaptive reactions lower myocyte number and contractile responses, hibernating cardiac myocytes can be saved by restoring blood flow, as it reprograms cell protein expression to normalize metabolism and contractile activity (Depre & Vatner, 2005; Slezak et al., 2009). The tissue cells' response to ischemia and cell survival is affected by the severity and time period of ischemia and by pathologic events that are initiated via reperfusion. Thus, stunning may happen after a transient period of ischemia (5—20 minutes in the heart), with reperfusion resulting in cell dysfunction, followed by slow and retarded recovery. Hibernation happens with a long time duration or repetitive intermittent decrease in the blood flow that are modest in degree. It is related to contractile disruption saved by revascularization. However, prolonged severe ischemia, which is later accompanied by reperfusion, leads to permanent impairment resulting in a decrease of functioning myocardium. The longer the time duration—and severe the period of ischemia—the greater the role in irreversible damage and cell death to postinfarct dysfunction (Fig. 1.2). However, the exact process by which reversible ischemia eventually turns into irreversible cell death is still controversial. But there is a significant chance that it is related to contemporaneous loss of a critical amount of ATP, generation of ROS, metabolically and mechanically induced membrane and cytoskeletal damage, calcium overload, sodium pump failure, and opening of the MPT pore.

Organ-specific susceptibility to ischemia/reperfusion

One of the most notable findings made in experimental models of I/R is that the damage response following reperfusion is correlated with the time period of ischemia (Bulkley, 1987). Therefore restoration of blood supply to the damaged organ as soon as possible is highly important. There are common, fundamental characteristics of the reaction to I/R. They involve the ROS, cytokines and chemokines release from activated endothelium and tissue-resident macrophages and mast cells; gathering of, activation, and endothelial adhesion/emigration accumulation of neutrophils and other formed elements in the blood; endothelial dysfunction; and paren-chymal damage. However, organ-specific variations affect the amount, intensity, and reversibility of organ injury. The biological basis for these variations is not well explained. In all tissues, cooling reduces cellular damage, which can be used intraoperatively or for improved preservation of organ transplants during transport (Baumgartner et al., 1989). Permanent injury is identifiable at less than 20 minutes of ischemia (Ordy et al., 1993). The brain is the most sensitive organ to low blood supply. Clinically there is a high possibility of focal cerebral ischemia (termed ischemic stroke). It erupts as a localized decrease in regional blood supply in a specific vascular territory as a consequence of thromboembolic or atherothrombotic vasoocclusive pathology. Although damage occurs rap-idly, the actual time window for therapeutic intervention is longer because not all cells are affected to the same extent after a given time period of ischemia. Hence, optimal results are seen if thrombolytic ther-apy is started within the first 90 minutes after the beginning of symptoms (Hacke et al., 2004). However, notable improvements in clinical results can still be obtained if the blood supply is reestablished within 3 hours (or within 4.5 hours in particular patient populations) (Bluhmki et al., 2009).

A plethora of unique features of the brain plays a role in its sensitivity: The brain is responsible for 20%—25% of total body oxygen utilization, having the highest metabolic activity per unit weight of any organ (Kristian, 2004; Lee, Grabb, Zipfel, & Choi, 2000). This high metabolic requirement results in an absolute demand for glucose as an energy sub-strate, but with low levels of stored glucose/glycogen as compared to other tissues (Kristian, 2004; Lee et al., 2000). Unlike the brain, muscle has the potential of having limited periods of anaerobic metabolism, and both muscle and liver have comparatively bigger stores of carbohydrate. The brain has comparatively low levels of protective antioxidant activities, for instance, superoxide dismutase, catalase, glutathione peroxidase (Adibhatla & Hatcher, 2010) and heme oxygenase-1 (Damle et al., 2009) than heart, liver, kidney, and lung. The brain also has lower levels of cytochrome c oxidase (Adibhatla & Hatcher, 2010), which are expected to result in lower ATP production and higher superoxide release from the mitochondrial electron transport chain. The brain has higher levels of polyunsaturated fatty acids, which have a high tendency to oxidative damage (Adibhatla & Hatcher, 2010). I/R can elicit excessive release of certain neurotransmitters, for example, glutamate and dopamine (Lee et al., 2000), which upon debasement of these neurotransmitters' postreceptor signaling pathways gives rise to neuronal calcium overload and subsequent cytotoxicity.

In the heart, the situation is similar but the therapeutic window is slightly longer. Irreversible cardiomyocyte damage occurs after about 20 minutes of ischemia. Like the brain, the sooner blood flow is restored, the better are survival rates and restoration of functioning myocardium. Intervention within the first 2 hours is best (Boersma, Maas, Deckers, & Simoons, 1996), but even after 12 hours of ischemia, reopening of the respective coronary arteries improves the outcome (LATE Study Group, 1993). In heart, mast cells and infiltrating fibroblasts evoke fibrosis (Frangogiannis, 2008; Willems, Havenith, De Mey, & Daemen, 1994). The actual role of mast cells is unclear, but the fibroblasts transdifferentiate and proliferate as myofibroblasts, and secrete collagen and other matrix proteins. An excess of this causes fibrosis and impairment of cardiac function. On the contrary, postischemic brain damage does not cause fibrosis, but instead it leads to glial cell activation (Dirnagl, Iadecola, & Moskowitz, 1999) and proteolysis of extracellular matrix, especially basal lamina, by enzymes called matrix metalloproteases. This results in astrocyte and endothelial detachment from basal lamina along with an increase in brain microvascular permeability, as well as glial and endothelial apoptosis (Winquist & Kerr, 1997).

The next most vulnerable organ is the kidney. By means of open renal surgery, it has been confirmed that no permanent organ damage occurs after normothermic ischemia of 30 minutes or less (McDougal, 1988). In animal models, even longer occlusions of the renal vessels appear to be feasible (Humphreys et al., 2009). In renal parenchyma, oxygenation is graded with the highest oxygen levels noted in the cortex, medium levels in the outer medulla, and the lowest levels in the papillae. As a consequence, cortical cells are the most vulnerable to ischemia, while cells in the outer medulla are less sensitive to a hypoxic environment as they can shift to oxygen-independent metabolism. Inner medullary and papillae cells use mostly glucose to generate ATP via anaerobic glycolysis. Hence, these regions demonstrate reduced sensitivity to ischemia.

While the "point of no return" is fairly easy to estimate in brain, heart, and kidney, the time window for successful intervention is much harder to assess in the case of intestinal ischemia. On the other hand, clinical symptoms are initially often subtle, making it impossible to recognize the onset of ischemia. If the diagnosis is made within 24 hours of the onset of symptoms and prompt treatment initiated, acute mesenteric ischemia has about a 50% survival rate, whereas this rate drops to 30% or less with a delayed diagnosis (Kassahun, Schulz, Richter, & Hauss, 2008). In experimental models, it has been shown that the degree of mucosal damage is a direct function of time elapsed from the onset of mesenteric artery occlusion. The first histological changes occur after 30 minutes and more eminent destruction of the villi is seen after 60 minutes (Ikeda et al., 1998). After revascularization, mucosal regeneration via cell migration occurs quickly even after 90 minutes of ischemia (Park & Haglund, 1992). It is essential to pinpoint that occlusion of the superior mesenteric artery (SMA) produces a gradient of ischemia along the bowel, with the severity of ischemia being greatest in distal small intestine and proximal colon, leaving the middle and distal colon unaffected (Premen, Banchs, Womack, Kvietys, & Granger, 1987). Furthermore, the ischemia is localized to the mucosal and submucosal layers of the bowel, while the muscularis/serosa remains unaffected. Collateral perfusion maintains minimal blood flow to the total intestinal wall but is more effective in supplying the muscularis/serosa than the mucosa/submucosa after SMA occlusion. Whereas complete SMA occlusion completely stops jejunal, ileal, and colonic blood flow in neonates (1 day to 1 month old), such observations may have important implications for the pathogenesis of neonatal necrotizing enterocolitis (Granger, 1988).

Intestinal I/R is accompanied by an increased luminal epithelial permeability and admittance of bacterial molecules (e.g., enterotoxin). The presence of bacteria themselves can result in sepsis and multiple organ failure, if the degree of ischemia is severe or the volume of ischemic mesenteric tissue is large (Kinross et al., 2009; Souza et al., 2004). Indeed, germ-free mice exhibit decreased local (intestinal) and remote (lung) injury following mesenteric I/R compared to its effects on conventional mice, because they were associated with decreased expression of proinflammatory cytokines and neutrophil sequestration (Souza et al., 2004). The lack of commensal flora is also due to increased expression of IL-10, an antiinflammatory cytokine. Antibodies directed against IL-10 that retard their function reverse the protection against I/R-induced inflammation and injury in germ-free mice. Similarly protection is found in germ-free mice subjected to hemorrhagic shock (Ferraro et al., 1995). Not long ago, these results were confirmed by work showing that diminishment of gut commensal bacteria by broad-spectrum antibiotic cocktail reduces intestinal I/R injury (Yoshiya et al., 2011) and lung injury induced by bowel ischemia (Sorkine et al., 1997).

Bacterial deletion decreases the expression of toll-like receptor2 and TLR4, well-known receptors for gram-positive and gram-negative bacteria (Yoshiya et al., 2011). Because of that, there is reduced expression of proinflammatory mediators (Tumor necrosis factor (TNF), Interleukin-6 (IL-6), and Cyclooygenase 2 (COX-2)), reduced complement and immunoglobulin deposition, and B-lymphocyte recruitment. Interestingly, probiotic colonizing of the intestine by oral administration of *Lactobacillus plantarum* for 2 weeks reduced bacterial translocation to extraintestinal areas, decreased the origination of proinflammatory cytokines, and reduced epithelial apoptosis and disruption of the mucosa induced by mesenteric I/R, relative to conventional animals (Wang, Huang, Zhang, Li, & Li, 2011). The results of these studies clearly indicate that the intestinal microflora plays a critical role in local and remote injury following gut I/R. These effects may be modulated by altering constituent commensal bacteria populations (Alverdy & Chang, 2008; Kinross et al., 2009; Kinross, Darzi, & Nicholson, 2011; Weber & Noels, 2011). These results were recently applied to myocardial I/R, where intestinal dysbiosis induced by vancomycin treatment prior to induction of coronary occlusion resulted in smaller infarcts, improved postischemic recovery of mechanical function, and decreased circulating leptin levels. These protective effects were replicated in animals fed a probiotic product containing *L. plantarum* (Lam et al., 2012).

Ischemia of skeletal muscle is much better tolerated. We know that acute arterial injuries may require emergency application of tourniquets Similarly, it is well known that hours of limb ischemia are well tolerated with good results obtained if the tourniquet is briefly released after the first 1.5—2 hours (Sapega, Heppenstall, Chance, Park, & Sokolow, 1985). Furthermore, skeletal muscle can regenerate even after extensive injury (Wagers & Conboy, 2005).

Tissues that contain very little or no vasculature are barely affected by ischemia. For example, cornea transplants can be stored in tissue culture media for 3 weeks with only little damage to endothelial cells (Smith & Johnson, 2010).

Since the microvasculature is the first site where recruitment of inflammatory cells takes place, tissue differences in the structure and function of microvascular beds are likely to play an important role in tissue responses to I/R. The generally accepted paradigm of endothelial selectin-dependent rolling and integrin-dependent adhesion of leukocytes during inflammation was developed using studies of these mechanisms in microvascular beds that are readily envisioned via intravital microscopy—for example, those of the intestinal mesentery and cremaster muscle (Granger & Korthuis, 1995). It might be expected that significant structural and functional differences between the microvasculature of different tissues should be correlated with similar differences in inflammatory processes The leukocyte—endothelial recruitment paradigm described in mesentery and cremaster seems to be enough to explain the process in most other tissues, except the liver, where the role of selectins appears to depend upon which microvascular bed is examined. Sinusoids, carrying blood from portal venules and hepatic arterioles, do not express P- and E-selectins nor support selectin-mediated rolling, whereas postsinusoidal venules do (Liu & Kubes, 2003). The role of integrins in hepatic sinusoids has also been questioned (Liu & Kubes, 2003). In this particular microvascular bed, leukocyte accumulation may be more effected by physical factors such as a vessel diameter close to that of leukocytes themselves (Liu & Kubes, 2003). Another "peculiar" organ, from the perspective of leukocyte adhesion and migration, is the lung. Unlike other tissues, where polymorphonuclear (PMN) adhesion and migration take place in relatively large postcapillary venules, in lung this process occurs primarily in alveolar capillaries that have a smaller diameter (Burns, Smith, & Walker, 2003). Moreover, the primacy of neutrophils themselves in ischemia reperfusion injury (IRI) is not obvious in some tissues. For example, lymphocytes and monocytes may play a more important role in mediating injury responses in kidney (Jang, Ko, Wasowska, & Rabb, 2009) and brain (Yilmaz & Granger, 2008).

The response to neonatal hearts to I/R is controversial. Some studies indicate that neonatal hearts are more vulnerable to ischemia (Wittnich, 1992), whereas other studies demonstrate enhanced tolerance of the immature myocardium (Julia, Kofsky, Buckberg, Young, & Bugyi, 1990).

It is quite likely that differences in species chosen for the studies are responsible for such discrepancies. Again, using the heart as an example, direct species comparisons using identical protocols for I/R have shown that isolated hearts from rabbits, hamsters, ferrets, gerbils, rats, mice, and guinea pigs differ substantially in injury susceptibility (Galinanes & Hearse, 1990). Furthermore, even within the same species, some strains are rather resistant to ischemia, whereas others are prone to injury. Examples can be found in studies of the heart (Barnabei, Palpant, & Metzger, 2010), brain (Barone, Knudsen, Nelson, Feuerstein, & Willette, 1993), kidney (Burne, Haq, Matsuse, Mohapatra, & Rabb, 2000), and lung (Dodd-o, Hristopoulos, Welsh-Servinsky, Tankersley, & Pearse, 2006), indicating that this presents a serious experimental limitation across all organ systems.

Remote organ injury

Effects of I/R are not necessarily limited to the tissue undergoing the initial ischemia. That is, a frequent consequence induced by reperfusion after localized tissue ischemia is injury to other organ systems, so-called distant or remote organ injury (ROI). This phenomenon can arise from I/R in most tissues, including gut (Carden & Granger, 2000; He, Dong, Chen, Zhou, & Shu, 2011; Santora et al., 2010; Sorkine et al., 1997), lung (Esme, Fidan, Koken, & Solak, 2006), liver (Hirsch et al., 2008), kidney (Vaghasiya, Sheth, Bhalodia, & Jivani, 2010), skeletal muscle (Vega et al., 2000), and heart (Barry et al., 1997). The ultimate expression of ROI is multiple organ dysfunction syndrome, resulting from I/R in gut, liver, skeletal muscle, aortic surgery involving occlusion—reperfusion, and circulatory shock (Carden & Granger, 2000; Santora et al., 2010). In this regard, the lungs (being the first major capillary bed exposed to postischemic blood) are especially vulnerable, particularly after I/R of the gut and/or liver (Carden & Granger, 2000; He et al., 2011). Indeed, one of the first clinical symptoms preceding multiple organ failure is respiratory dysfunction (Carden & Granger, 2000; He et al., 2011; Santora et al., 2010).

Examination of the mechanisms underlying ROI has found roles for the same factors implicated in the local organ dysfunction produced by IRI: ROS, leukocytes, and inflammatory mediators. A common finding has been that one or more circulating factors are responsible for the effect on organs distant from the one undergoing the initial insult (Carden & Granger, 2000; He et al., 2011; Santora et al., 2010). These factors may be directly released from the primary injured tissue or indirectly from activated leukocytes or other inflammatory cells.

Xanthine oxidase (XO), which creates superoxide and hydrogen peroxide, has been implicated as an important factor in ROI in liver, lung, and cardiac muscle after gut I/R. The mechanism for XO-mediated systemic effects is not clear, but it may involve the following processes: generation of high amounts of ROS by circulating enzyme, close association of XO with endothelial cell surface and consequent high local ROS concentrations, or XO-derived oxidant-induced release of chemotactic factors that can promote recruitment of PMN to organs distant from the initial injury.

Just as the primary organ subjected to I/R, inflammatory leukocytes play a major role in injury to remote organs. A key event appears to be the activation or "priming" of PMN in a postischemic vascular bed, followed by recruitment of the activated PMNs to remote tissues (Carden & Granger, 2000). This involves the activation of PMN and also of endothelial cells in distant tissues, characterized by increased surface expression of endothelial adhesion molecules. Systemic release of inflammatory mediators from the primary injured tissue and/or from recruited monocytes and neutrophils as well as systemic complement activation has all been reported to promote systemic activation of and recruitment of PMN to sites distant from initial I/R. If intestinal I/R is involved in ROI, bacteria can cross the mucosal barrier, causing systemic infection and sepsis (Stallion et al., 2005). In recent years, it is considered that ischemic-reperfused intestine releases cytokines and other inflammatory mediators into the intestinal lymph; these agents enter the systemic circulation at the thoracic duct (Deitch, 2010; Deitch, Xu, & Kaise, 2006; He et al., 2011), and it has been shown that lymph is their primary route of entry, since ligation of the mesenteric lymph duct can prevent ROI (Deitch, 2010; Deitch et al., 2006).

Over the past decade, it has become evident that neurogenic signals contribute to inflammatory responses (Ahluwalia, De Felipe, O'Brien, Hunt, & Perretti, 1998; Bhatia et al., 1998; Bozic, Lu, Hopken, Gerard, & Gerard, 1996; Cao et al., 2000; Souza, Mendonca, de, Poole, & Teixeira, 2002), including ROI. The proinflammatory phenotype produced by intestinal I/R can be significantly reduced by treating with the sensory nerve toxin, capsaicin, and tachykinin receptor antagonists (Souza et al., 2002). Significantly, the above-written protection is observed both in the gut and in the lung (Souza et al., 2002), enhancing the potential importance of neurogenic signals in ROI. The most likely mediators for these effects are neuropeptides released from both sensory nerve endings and inflammatory cells (Quartara & Maggi, 1998). It has been proposed that neurokinin-dependent signaling may contribute to either or both initiation of I/R-induced inflammatory responses via initial release of lipid mediators such as PAF, or amplification of an already present inflammatory phenotype (Souza et al., 2002).

Pathophysiology of ischemia/reperfusion injury

After the occlusion of an epicardial coronary artery, the myocardium previously perfused by the occluded artery is in jeopardy. The hypoperfused myocardial zone during myocardial infarction is known as the area at risk (AAR). If the coronary artery is not rapidly reperfused and no collateral circulation is present, most of the AAR becomes necrotic. Given that many patients receive timely reperfusion therapy, part of the AAR remains free of necrosis: the so-called salvaged myocardium. The typical morphological features of reperfused myocardial infarction are contraction bands, karyolysis, mitochondrial swelling and disruption, and membrane disruption in cardiomyocytes, microvascular destruction, interstitial hemorrhage, and inflammation (Kloner et al., 1980; Reimer, Jennings, & Tatum, 1983). Experimental studies identified the determinants of myocardial infarct size as: (1) the size of the AAR (Miura, Yellon, Hearse, & Downey, 1987); (2) the duration of myocardial ischemia (Reimer & Jennings, 1979; Reimer, Lowe, Rasmussen, & Jennings, 1977); (3) the amount of residual blood flow through collaterals (Miura et al., 1987; Reimer & Jennings, 1979); (4) the temperature of the tissue during ischemia; and (5) the hemodynamic situation during ischemia (Duncker et al., 1996). The most notable hemodynamic parameter is heart rate, which determines both myocardial demand and coronary blood flow (Heusch, 2008); however, hemodynamics influence infarct size only to a limited degree. Infarct size is largely determined by lack of blood/energy supply and less by myocardial demand, which is significantly reduced by the regional lack of contraction (Skyschally, Schulz, & Heusch, 2008).

The seminal studies by Maroko et al. (1972) and Ginks et al. (1972) 40 years ago first demonstrated that reperfusion salvages myocardium from infarction, and these studies initiated the ongoing success story of reperfusion therapy (Van de Werf, 2014). The potential of reperfusion to induce additional injury secondary to the ischemic damage emerged soon afterward with the identification of stunning as a reversible form of myocardial reperfusion injury (Braunwald & Kloner, 1985). Although the contribution of reperfusion injury to final infarct size has been disputed in the past, today it is accepted that reperfusion can induce additional damage to the myocardium. This view is supported by strong evidence that interventions applied at the end of the ischemic period (i.e., coinciding with reperfusion) can reduce infarct size. It was already recognized in the mid-1980s that gentle reperfusion at low pressure results in significantly less edema and a smaller infarct size than standard abrupt reperfusion at normal pressure (Okamoto, Allen, Buckberg, Bugyi, & Leaf, 1986). This idea was later developed by Zhao et al. (2003), who demonstrated reduction of infarct size by brief episodes of coronary reocclusion/reflow at the time of reperfusion, a strategy called ischemic postconditioning (Heusch, 2004). Because these interventions are applied at the end of the ischemic period, they cannot reduce infarct size by reducing ischemic damage and, thus, must reduce reperfusion-related damage. From these observations it is clear that reperfusion injury contributes to infarct size, and all conditioning strategies that protect the myocardium and reduce infarct size act only in conjunction with eventual reperfusion (Heusch et al., 2014).

Role of microcirculation in infarct size after ST-segment elevation

The myocardial I/R is a complex phenomenon where the role of microcirculation is very important. It is the interface between the epicardial vessel and the cardiomyocytes. Thus, it is not important how efficiently and quickly the blood flow is restored to the epicardial artery; if a microvascular obstruction (MVO) is present, the myocardial tissue will persist without efficient perfusion. MVO (also known as the no-reflow phenomenon) during I/R greatly affects the final infarct size and causes morbidity/mortality without being dependent on any other factor (Eitel et al., 2014). Kloner et al. was the first to explain the no-flow phenomenon (Kloner, Ganote, & Jennings, 1974) in dogs that were exposed to 90 minutes of coronary occlusion and subsequent reperfusion. Severe capillary damage was seen with significant swelling and rupture of endothelial cells, and intraluminal thrombosis, swollen and permanently damaged cardiomyocytes were observed around it. Clinically, the no-reflow phenomenon is seen in 10%—30% of patients with reperfused ST-segment elevation (STEMI) (Bekkers, Yazdani, Virmani, & Waltenberger, 2010;

Niccoli, Burzotta, Galiuto, & Crea, 2009), although successful recanalization of the epicardial coronary arteries has been performed. MVO in such people is detected angiographically from slow or no reflow of contrast medium or by cardiac magnetic resonance (CMR). MVO forms after a few minutes of starting reperfusion (Ambrosio, Weisman, Mannisi, & Becker, 1989; Reffelmann & Kloner, 2002) and remains for at least 1 week (Mewton et al., 2013; Rochitte et al., 1998). MVO is often limited to the infarcted myocardium (Kloner et al., 1974), however the presence of no-reflow phenomena within the AAR, but outside the infarcted area, has systematically included. MVO affects the washout of reduction equivalents and dehydrogenases that is a requirement for valid delineation of infarcted tissue by triphenyltetrazolium chloride (TTC) staining, thus resulting infarct size being underestimated. Decreased coronary blood flow is also seen outside of the AAR, but this does not reflect a no-reflow phenomenon, but it is the consequence of reflex-mediated alpha-adrenergic coronary vasoconstriction (Gregorini et al., 1999; Gregorini, Marco, & Heusch, 2012; Heusch et al., 2000).

Several mechanisms have been put proposed as factors leading to MVO: (1) embolization of particulate debris from the damaged culprit atherosclerotic lesion, with physical obstruction/occlusion of the coronary microcirculation (Heusch et al., 2009); (2) platelet and platelet/leukocyte aggregates that are released from the site of the culprit lesion, form in the coronary microcirculation, or come with the blood supply, where they contribute to the general inflammatory status associated with STEMI (Barrabes et al., 2010); (3) soluble vasoconstrictor substances released from the culprit lesion cause severe vasoconstriction (Kleinbongard et al., 2011, 2013); (4) edema in the surrounding myocardium results in extravascular coronary microvascular compression (Manciet, Poole, McDonagh, Copeland, & Mathieu-Costello, 1994); and (5) primary physical damage to the capillary endothelium (Kloner et al., 1974). These mechanisms do not affect independently but they work with each other, and their individual effect on the disruption of myocardial reperfusion may change temporally and spatially. Permanent damage to cardiomyocytes and the coronary microcirculation are intimately connected (Heusch, Kleinbongard, & Skyschally, 2013). High intramyocardial pressure, with a predominant contribution from edema, might be the main reason for MVO in the endocardial layer, while microembolization might cause spreading of infarct in the border zone (Skyschally, Walter, & Heusch, 2013). Although it is debatable that lack of efficient tissue perfusion in areas of MVO will maintain the muscle ischemia and consequently result in infarct expansion, there is presently no proof to explain a causal role for microvascular coronary obstruction in myocardial infarction.

The contribution of leukocytes in the formation of myocardial infarct is contentious, and they may have a more significant role in infarct healing and remodeling instead of estimating the infarct size (Baxter, 2002). However, the effect of intravascular leukocytes to MVO and infarct size needs more study. Myocardial infarction and MVO presently appear to be parallel phenomena that are caused by not very different mechanisms: a primary energetic lack and the resulting excessive development of ROS upon reperfusion. It is obvious that postconditioning decreases myocardial infarct size and MVO (Heusch, 2004; Zhao et al., 2003).

Cardiomyocyte necrosis, apoptosis, autophagy, and necroptosis

Myocardial infarction has earlier been seen as a presentation of necrotic cell death, but currently, various forms of cardiomyocyte death have been observed during I/R and are considered to affect the ultimate infarct size.

Necrosis is seen as myofibrillar contraction bands, swollen and damaged mitochondria, disruption of membranes of cardiomyocyte, impairment of microvascular, hemorrhage, and inflammation. Most of these morphological features are aggravated by reperfusion (Kloner et al., 1980; Kloner, Ellis, Lange, & Braunwald, 1983; Roberts, Schoen, & Kloner, 1983). Necrosis is a consequence of unregulated and uncoordinated pathophysiological processes. During ischemia, the acidosis due to anaerobic glycolysis leads to more influx of Na^+ through the Na^+/H^+-exchanger, and intracellular Na^+ collection is enhanced by the inhibition of Na^+/K^+-ATPase because of the unavailability of ATPs (Ladilov, Siegmund, & Piper, 1995; Tani & Neely, 1989). This may lead to calcium overload due to reverse mode operation of Na^+/Ca^{++} exchanger. When reperfusion is established, the rapid normalization of pH and reenergization in the context of elevated cytosolic Ca^{++} causes oscillatory release and reabsorption of Ca^{++} into the sarcoplasmic reticulum, leading to uncontrolled increased myofibrillar hypercontraction (Ladilov et al., 1995; Piper, Abdallah, & Schafer, 2004; Piper, Meuter, & Schafer, 2003). The normalization of the acidic pH also stimulates calpain, which metabolizes the cytoskeleton and the sarcolemma (Inserte, Hernando, & Garcia-Dorado, 2012). The increased cytosolic concentrations of Na^+ and Ca^{++}

lead to intracellular edema when extracellular osmolarity is quickly reduced to normal by reperfusion. Eventually, a high amount of ROS causes sarcolemmal injury (Schluter, Jakob, Ruiz-Meana, Garcia-Dorado, & Piper, 1996). Inflammation occurs after necrosis.

Different from necrosis, apoptosis, autophagy, and necroptosis are programmed processes with specific underlying signal transduction mechanisms (Chiong et al., 2011; Orogo & Gustafsson, 2013). Apoptosis is an energy-consuming type of cell death having characteristic deoxyribonucleic acid strand breaks that are detected by deoxyribonucleic acid laddering and/or terminal deoxynucleotidyl transferase dUTP nick-end labeling staining (Zhang & Xu, 2000). Apoptosis starts extrinsically by stimulating sarcolemmal receptors, especially Fas Cell surface death receptor (FAS) and tumor necrosis factor α receptors (Kleinbongard, Schulz, & Heusch, 2011), and intrinsically by mitochondrial release of cytochrome c, which starts caspase activation causing intracellular proteolysis, in the absence of inflammation (Orogo & Gustafsson, 2013). The important part is when the mitochondrial permeability transition pore (MPTP) opens. It is a very important channel that normally does not open unless high levels of intracellular calcium, inorganic phosphate and ROS usually found in I/R injury (Kroemer & Reed, 2000; Weiss, Korge, Honda, & Ping, 2003). Consequently, mitochondrial matrix swelling occurs, ultimately harming the outer membrane and releasing cytochrome c to the cytosol, where it leads to the stimulation of the caspase cascade. Proapoptotic and antiapoptotic proteins of the B cell lymphoma (BCL) family interact with the MPTP (Baines, 2009). Here, the earlier aspect of the MPTP has been debated, as all of its purported constituents are dispensable under some conditions, the MPTP might form from F-ATP synthase (Carraro et al., 2014).

Autophagy is a process that degrades proteins using lysosomes and recycles, along with proteins of mitochondria (mitophagy) (Dong, Undyala, Gottlieb, Mentzer, & Przyklenk, 2010). Although autophagy promotes cell death but its not harmful (Przyklenk et al., 2011). For example, in pigs exposed to 45 minutes of coronary occlusion and reperfusion, the purported autophagy inducer chloramphenicol decreased infarct size (Sala-Mercado et al., 2010). However, the effect of autophagy in myocardial I/R damage in humans is debatable (Gedik et al., 2014; Singh et al., 2014).

Necroptosis has properties of both necrosis and apoptosis, but is markedly regulated by activating receptor-interacting protein kinases

1 and 3 (Zhou & Yuan, 2014) and can be inhibited by substances such as necrostatin (Oerlemans et al., 2013).

It is currently not clear as to what extent necrosis, apoptosis, autophagy, and necroptosis are mutually exclusive processes and how much exactly each leads to infarct size. Typical characteristics of apoptosis (terminal deoxynucleotidyl transferase dUTP nick-end labeling staining) and autophagy (characteristic protein expression) are present in the TTC staining-defined infarct zone, which was earlier considered as necrotic. The opening of the MPTP seems to be trigger for necrosis, apoptosis, and necroptosis, and mitochondria are also decisive in mitophagy/autophagy. The significance of regulated types of cardiomyocyte cell death in I/R injury is more associated with their specific signal transduction processes. The possibility is suggested by the recognition of the various modes of cardiomyocyte death during infarction.

Therapeutic advancement during last decade

Sudden occlusion of an epicardial coronary artery leads to acute myocardial infarction expressed as STEMI. Consequently, ischemia is observed in the myocardium far from the occlusion site. Irreversible injury occurs due to unrelieved ischemia to the myocardium earlier supplied by the occluded artery. Myocardium is damaged and replaced by fibrous scar tissue. Since scar tissue does not lead to myocardial contractile function, if the scar is big, global left ventricular (LV) contractile function is disrupted, causing progressive chronic heart failure. After coronary thrombosis as the reason (not the consequence) of STEMI in most cases, timely reestablishment of blood flow to the ischemic myocardium (reperfusion) need to be considered as the standard treatment in these cases. Immediate reperfusion demonstrate attenuation in infarct size, having a positive effect on longterm myocardial function, altering the healing pattern of the infarcted zone and decreasing mortality. A lot of experimental and clinical studies shown myocardial injury associated with this phenomena, called reperfusion injury. Consequently, injury to the myocardium during an STEMI is better explained as I/R injury, the effect of the ischemic and reperfusion mechanism.

Considering above mentioned issues, it is obvious that I/R is a complex phenomena that may lead to impairment of the myocardial function, as following explained by Ibanez et al. (Ibáñez, Heusch, Ovize, & Van de Werf, 2015).

1. The initial significant involvement is the epicardial artery. Atherosclerotic plaque disruption with superimposed thrombus causes an abrupt stop of oxygen and nutrient supply away from the site of blockage. The opening of the epicardial vessel by mechanical or pharmacological means, along with the decrease in thrombus burden by adjuvant antiplatelet/anticoagulant therapies, is just the first step for healing the myocardium. During the reperfusion (either mechanical by primary angioplasty or pharmacological by thrombolytics), thrombus material and other plaque debris can be distally embolized leading to MVO.

2. Circulating cells harm the myocardium: activated platelet and leukocytes in the bloodstream cause the formation of thrombus, and plugins that can embolize distally into the microcirculation upon resting blood flow across the culprit lesion (a process not dependent on plaque debris microembolization).

3. The microcirculation (net of capillaries) is an important factor in the final stage of myocardium during I/R. Once the epicardial vessel flow is reestablished, efficient tissue perfusion is mostly controlled by the microcirculation. Plaque debris and platelet/neutrophil accumulation might lead to a mechanical blockage of the microcirculation precluding efficient tissue perfusion. The tissue edema formation following reperfusion can cause external compression of the microcirculation, decreasing the perfusion capacity of the capillary network (double arrows). Eventually, the rupture of microcirculation can occur because of the earlier injury and permit the leakiness of circulating cells into the interstitial space.

4. Red blood cell deposits (hemorrhage) are very dangerous as the release of iron, causes inflammation.

5. Cardiomyocytes that have survived the ischemic phase suffer during the reperfusion period due to several intracellular pathways triggered at reperfusion (see text for detailed information about these processes).

6. After the whole I/R insult has passed, the significant infiltration of myocardial tissue by inflammatory cells can cause more injury to the myocardium.

Myocardial infraction (ST-segment elevation)

Because of the increasing use of preventive therapies and more efficient control of risk factors, the incidence of STEMI in Western countries has reduced in recent decades (O'Gara et al., 2013). Despite the progressive and slow reduction in its incidence, STEMI continues to be an important health issue, being reported as the leading cause of mortality/morbidity worldwide (O'Gara et al., 2013). As detailed later in this chapter, the implementation of timely reperfusion has led to a great decrease in the acute mortality related to STEMI. Risk-adjusted, in-hospital mortality has reduced from $\approx 20\%$ in the late 1980s to $\approx 5\%$ among STEMI patients treated in routine practice in 2008 (Eapen et al., 2012), reaching a plateau thereafter (Menees et al., 2013). However, these impressive reductions in mortality rates, due to the extensive use of reperfusion strategies and adjuvant pharmacological therapies, have led to a higher incidence of chronic heart failure. Although this outcome might initially appear paradoxical, the explanation is simple: patients with a significantly depressed cardiac function would not have gotten through the acute STEMI phase earlier, but with the advent of reperfusion, they now survive the index episode and live with a severely impaired heart (Eapen et al., 2012). In fact, STEMI is one of the leading causes of chronic heart failure. Postinfarction decreased left ventricular ejection fraction (LVEF) is one of the main reasons behind chronic heart failure globally (Callender et al., 2014). The favorable results of reperfusion have led to a paradigm shift in clinical research in the field of STEMI: focus is not only to lower the mortality rate anymore (already very low), but it is to tackle the result of better survival: postinfarction heart failure.

Successful clinical research has led to treatments for chronic heart failure (drugs and devices) that limit long-term mortality in STEMI survivors with low LVEF (Yancy et al., 2013). However, these strategies are expensive, precluding their universal usage (LaPointe et al., 2011). Chronic treatment of heart failure is associated with a huge socioeconomic burden on people and healthcare systems. As explained later in this chapter, infarct size is the main determinant of negative postinfarction results, including heart failure (Larose et al., 2010). Therapies that decrease infarct size are immediately sought under the hypothesis that smaller infarctions will have an improved long-term heart performance and that this will turn into lesser harmful clinical events (Heusch, 2013; Kloner, 2013).

Pathophysiology of ischemia/reperfusion injury

After the blockage of an epicardial coronary artery, the myocardium earlier perfused by the occluded artery is in danger. The AAR is the hypoperfused myocardial zone during myocardial infarction. Failure to quickly reperfuse the coronary artery, and the absence of collateral circulation, leads to most of the AAR becoming necrotic. These days mostly patients may reach at healthcare facility timely and in no time receive possible reperfusion therapy, part of the AAR does not become necrotic: known as the so-called salvaged myocardium. Contraction bands, karyolysis, swelling of mitochondria, and membrane disruption in cardiomyocytes, along with microvascular damage, interstitial hemorrhage, and inflammatory response are the typical morphological features of reperfused myocardial infarction (Kloner et al., 1980; Reimer et al., 1983). Experimental studies showed the factors that affect the size of myocardial infarct as (1) the size of the AAR (Miura et al., 1987); (2) the duration of myocardial ischemia (Reimer & Jennings, 1979; Reimer et al., 1977); (3) the amount of residual blood flow through collaterals (Miura et al., 1987; Reimer & Jennings, 1979); (4) the temperature of the tissue during ischemia; and (5) the hemodynamic situation during ischemia (Duncker et al., 1996). Heart rate is the most significant hemodynamic factor, which shows the myocardial demand, and coronary blood flow (Heusch, 2008); however, hemodynamics affect infarct size only to an extent, and infarct size is, thus, mainly affected by the absence of blood flow and less by myocardial demand, which is greatly decreased by the local absence of contraction (Skyschally et al., 2008).

Four decades ago, the seminal studies by Maroko et al. (1972) and Ginks et al. (1972) explained that reperfusion salvages myocardium from infarction, and these studies started the positive results of reperfusion therapy (Van de Werf, 2014). The property of reperfusion to cause more harm secondary to the ischemic damage was seen immediately after stunning was considered as a reversible type of myocardial reperfusion injury (Braunwald & Kloner, 1985). Although the role of reperfusion injury to the ultimate infarct size was debated earlier, here it is agreed upon that reperfusion can be more harmful to the myocardium. This view is backed by a valid proof that treatment given at the end of the ischemic period (i.e., coinciding with reperfusion) decreases infarct size. It was established in the mid-1980s that gentle reperfusion at low pressure leads to greatly reduced edema and a smaller infarct size than standard sudden reperfusion at normal pressure (Okamoto et al., 1986). In early 21st century, the idea came by Zhao et al. (2003) and Heusch et al. (2004), and experimental work endorsed their idea about brief episodes of ischemia during reperfusion, a process known as Ischemia postconditioning. Since these treatments are used at the end of the ischemic period, they are unable to decrease the infarct size by limiting ischemic injury and, thus, must decrease reperfusion-related harmful effects. Therefore it is evident that reperfusion injury affects infarct size, and that all conditioning strategies that are protective for the myocardium and decrease infarct size work only along with final reperfusion (Heusch et al., 2014).

Preclinical models of myocardial ischemia/reperfusion injury

Why do we need pre-clinical models of ischemia reperfusion injury because patients suffering from I/R may have multiple risk factors, co-medications and co-morbidities (Ferdinandy, Hausenloy, Heusch, Baxter, & Schulz, 2014) probably, coronary circulation has functional and structural remodeling before STEMI (Heusch et al., 2012). There may or may not be formation of collateral circulation (Seiler, Stoller, Pitt, & Meier, 2013), past experience of prodromal angina attacks that might prove to have a protective effect by ischemic preconditioning (Heusch, 2001; Rezkalla & Kloner, 2004), or earlier coronary microembolization leading to patchy microinfarcts (Heusch et al., 2009). STEMI often starts by the sudden disruption of an atherosclerotic plaque and complete obstruction of an epicardial coronary artery; usually spontaneous or interventional reperfusion occurs after this, but occasionally in the absence of reperfusion (Skyschally et al., 2008). Since all of these factors lead to the ultimate infarct size, it is obvious that animal models cannot perfectly summarize a clinical situation. The majority of the information myocardial infarction is obtained from studies in anesthetized, young, healthy animals exposed to abrupt coronary blockage and reperfusion. Animal models are important in studying the pathophysiological aspect of humans, but the knowledge gained with a specific model can answer a limited number of questions. Hence, a superior animal model does not exist, each having its specific applications. Rodent models are useful in detecting potential mechanisms, but the conclusions cannot be applied to clinical environments because of their great anatomical/physiological variations from us. Pilot clinical trials are used only when valid and incontestable advantages are seen in large-animal models more similar to human anatomy and physiology. The past data on cardiovascular medicine is filled with unsuccessful clinical experience because this translational path was skipped.

Value of small animal models

Mice are extensively seen as subjects in experimental infarct experiments because they are easily bred, are cheap to use, and transgenic models are present. However, conclusions are greatly restricted due to their high heart rate (an order of magnitude more than humans) and the small heart size; therefore, during permanent coronary blockage, the inner myocardial layers are still supplied by oxygen and nutrients by diffusion. Hence, infarction in mice forms within 45–90 minutes (Redel et al., 2008; Thibault et al., 2007), but a maximum of 70% of the AAR is infarcted even with irreversible coronary obstruction (Michael et al., 1995). Rodents have higher heart rate than larger mammals; but where rats and rabbits have limited collateral blood flow and quick infarct progression (Manintveld et al., 2007), guinea pigs have sufficient collateral blood flow and extremely limited infarct progression (Schaper, Gorge, Winkler, & Schaper, 1988). Variations in the effect to myocardial infarction are observed in different mice strains. According to the genetic background, infarct size after the same procedure can differ by 30% (Guo et al., 2012).

Large-animal models of myocardial infarction

Large animals are mandatory before beginning human trials. Larger mammals include pigs and primates have reduced collateral blood flow, whereas cats and specifically dogs have a sufficient innate collateral circulation (Schaper et al., 1988). In pigs, infarction begins after 15–35 minutes after coronary occlusion and extends in a fashion that after 60–180 minutes, infarction is huge and occupies more than 80% of the AAR (Horneffer, Healy, Gott, & Gardner, 1987; Pich, Klein, Lindert, Nebendahl, & Kreuzer, 1988). Tolerance to I/R differs greatly across among pig strains. Primates exhibit a significant resistance to infarction, with less or no infarction seen after 40–60 minutes of coronary blockage, and even after 90 minutes of obstruction, the infarct size is less than in pigs (Yang et al., 2010). In dogs, infarction is mostly subendocardial after 40 minutes of coronary blockade and progresses to cover about 70% of the AAR after 6 hours, but even after irreversible obstruction a little zone of normal myocardium remains in the subepicardium [85] (Reimer et al., 1977). Given the different but notable collateral blood flow in dogs, infarct size is most accurately measured as a fraction of the AAR and by its inverse association with the residual blood flow (Skyschally et al.,

2008). Without resisting the confounders detailed earlier (age, comorbidities, comedication), infarct formation in large mammals is faster than in humans. From contrast-enhanced CMR analysis (Desmet et al. 2011; Hedström et al., 2009; Ibanez et al., 2013) and the amount of salvageable ischemic myocardium at the time of reperfusion (Piot et al., 2008; Staat et al., 2005), it can be calculated that around 30%—50% of the AAR is still functional after 4—6 hours from the initiation of symptoms of angina. Even after 12 hours of coronary obstruction, interventional reperfusion can highly restrict infarct size (Schömig et al., 2005). Yet, it is not well understood whether the faster infarct progression in larger mammals than in humans is associated with a formed collateral circulation similar to that in dogs (Seiler et al., 2013), a species-specific greater resistance to infarction as in primates (Yang et al., 2010), preinfarction anginal episodes that provide protection ischemic preconditioning (Rezkalla & Kloner, 2004), a little amount of short-term hibernation with contractile and metabolic adaptation to the decreased blood flow (Heusch & Schulz, 2002; Heusch, Schulz, & Rahimtoola, 2005), or background medication, significantly with platelet inhibitors (Roubille et al., 2012).

Ischemia—reperfusion injury has been ignored for a long time. Because of its important role in myocardial impairment resulting in bad results in various cardiovascular events, a lot of research is being carried out globally using multiple models/strategies. We hope in the future we will be able to protect the heart from I/R injury.

References

Adibhatla, R. M., & Hatcher, J. F. (2010). Lipid oxidation and peroxidation in CNS health and disease: From molecular mechanisms to therapeutic opportunities. *Antioxidants & Redox Signaling*, *12*(1), 125—169. Available from https://doi.org/10.1089/ars.2009.2668.

Ahluwalia, A., De Felipe, C., O'Brien, J., Hunt, S. P., & Perretti, M. (1998). Impaired IL-1beta-induced neutrophil accumulation in tachykinin NK1 receptor knockout mice. *British Journal of Pharmacology*, *124*(6), 1013—1015. Available from https://doi.org/10.1038/sj.bjp.0701978.

Alverdy, J. C., & Chang, E. B. (2008). The re-emerging role of the intestinal microflora in critical illness and inflammation: Why the gut hypothesis of sepsis syndrome will not go away. *Journal of Leukocyte Biology*, *83*(3), 461—466. Available from https://doi.org/10.1189/jlb.0607372.

Ambrosio, G., Weisman, H. F., Mannisi, J. A., & Becker, L. C. (1989). Progressive impairment of regional myocardial perfusion after initial restoration of postischemic blood flow. *Circulation*, *80*(6), 1846—1861.

Baines, C. P. (2009). The mitochondrial permeability transition pore and ischemia-reperfusion injury. *Basic Research in Cardiology*, *104*(2), 181—188. Available from https://doi.org/10.1007/s00395-009-0004-8.

Barnabei, M. S., Palpant, N. J., & Metzger, J. M. (2010). Influence of genetic background on ex vivo and in vivo cardiac function in several commonly used inbred mouse strains. *Physiological Genomics*, *42A*(2), 103—113. Available from https://doi.org/10.1152/physiolgenomics.00071.2010.

Barone, F. C., Knudsen, D. J., Nelson, A. H., Feuerstein, G. Z., & Willette, R. N. (1993). Mouse strain differences in susceptibility to cerebral ischemia are related to cerebral vascular anatomy. *Journal of Cerebral Blood Flow & Metabolism*, *13*(4), 683—692. Available from https://doi.org/10.1038/jcbfm.1993.87.

Barrabes, J. A., Inserte, J., Agullo, L., Alonso, A., Mirabet, M., & Garcia-Dorado, D. (2010). Microvascular thrombosis: An exciting but elusive therapeutic target in reperfused acute myocardial infarction. *Cardiovascular & Hematological Disorders-Drug Targets*, *10*(4), 273—283.

Barry, M. C., Wang, J. H., Kelly, C. J., Sheehan, S. J., Redmond, H. P., & Bouchier-Hayes, D. J. (1997). Plasma factors augment neutrophil and endothelial cell activation during aortic surgery. *European Journal of Vascular and Endovascular Surgery*, *13*(4), 381—387.

Baumgartner, W. A., Williams, G. M., Fraser, C. D., Jr., Cameron, D. E., Gardner, T. J., Burdick, J. F., ... Reitz, B. A. (1989). Cardiopulmonary bypass with profound hypothermia. An optimal preservation method for multiorgan procurement. *Transplantation*, *47*(1), 123—127.

Baxter, G. F. (2002). The neutrophil as a mediator of myocardial ischemia-reperfusion injury: Time to move on. *Basic Research in Cardiology*, *97*(4), 268—275. Available from https://doi.org/10.1007/s00395-002-0366-7.

Bekkers, S. C., Yazdani, S. K., Virmani, R., & Waltenberger, J. (2010). Microvascular obstruction: Underlying pathophysiology and clinical diagnosis. *Journal of American College Cardiology*, *55*(16), 1649—1660. Available from https://doi.org/10.1016/j.jacc.2009.12.037.

Bhatia, M., Saluja, A. K., Hofbauer, B., Frossard, J. L., Lee, H. S., Castagliuolo, I., ... Steer, M. L. (1998). Role of substance P and the neurokinin 1 receptor in acute pancreatitis and pancreatitis-associated lung injury. *Proceedings of the National Academy of Sciences of the United States of America*, *95*(8), 4760—4765.

Bluhmki, E., Chamorro, A., Davalos, A., Machnig, T., Sauce, C., Wahlgren, N., ... Hacke, W. (2009). Stroke treatment with alteplase given 3.0—4.5 h after onset of acute ischaemic stroke (ECASS III): Additional outcomes and subgroup analysis of a randomised controlled trial. *The Lancet Neurology*, *8*(12), 1095—1102. Available from https://doi.org/10.1016/s1474-4422(09)70264-9.

Boersma, E., Maas, A. C., Deckers, J. W., & Simoons, M. L. (1996). Early thrombolytic treatment in acute myocardial infarction: Reappraisal of the golden hour. *The Lancet*, *348*(9030), 771—775. Available from https://doi.org/10.1016/s0140-6736(96)02514-7.

Bolli, R., & Marban, E. (1999). Molecular and cellular mechanisms of myocardial stunning. *Physiological Reviews*, *79*(2), 609—634. Available from https://doi.org/10.1152/physrev.1999.79.2.609.

Bozic, C. R., Lu, B., Hopken, U. E., Gerard, C., & Gerard, N. P. (1996). Neurogenic amplification of immune complex inflammation. *Science, 273*(5282), 1722–1725.

Braunwald, E., & Kloner, R. A. (1985). Myocardial reperfusion: A double-edged sword?. *Journal of Clinical Investigation, 76*(5), 1713–1719. Available from https://doi.org/10.1172/jci112160.

Bulkley, G. B. (1987). Free radical-mediated reperfusion injury: A selective review. *The British Journal of Cancer Supplement, 8,* 66–73.

Burne, M. J., Haq, M., Matsuse, H., Mohapatra, S., & Rabb, H. (2000). Genetic susceptibility to renal ischemia reperfusion injury revealed in a murine model. *Transplantation, 69*(5), 1023–1025.

Burns, A. R., Smith, C. W., & Walker, D. C. (2003). Unique structural features that influence neutrophil emigration into the lung. *Physiological Reviews, 83*(2), 309–336. Available from https://doi.org/10.1152/physrev.00023.2002.

Callender, T., Woodward, M., Roth, G., Farzadfar, F., Lemarie, J. C., Gicquel, S., ... Rahimi, K. (2014). Heart failure care in low- and middle-income countries: A systematic review and meta-analysis. *PLoS Medicine, 11*(8), e1001699. Available from https://doi.org/10.1371/journal.pmed.1001699.

Cao, T., Pinter, E., Al-Rashed, S., Gerard, N., Hoult, J. R., & Brain, S. D. (2000). Neurokinin-1 receptor agonists are involved in mediating neutrophil accumulation in the inflamed, but not normal, cutaneous microvasculature: An in vivo study using neurokinin-1 receptor knockout mice. *Journal of Immunology, 164*(10), 5424–5429.

Carden, D. L., & Granger, D. N. (2000). Pathophysiology of ischaemia-reperfusion injury. *Journal of Pathology, 190*(3), 255–266. Available from https://doi.org/10.1002/(sici)1096-9896(200002)190:3 < 255::aid-path526 > 3.0.co;2-6.

Carraro, M., Giorgio, V., Sileikyte, J., Sartori, G., Forte, M., Lippe, G., & Bernardi, P. (2014). Channel formation by yeast F-ATP synthase and the role of dimerization in the mitochondrial permeability transition. *Journal of Biological Chemistry, 289*(23), 15980–15985. Available from https://doi.org/10.1074/jbc.C114.559633.

Chiong, M., Wang, Z. V., Pedrozo, Z., Cao, D. J., Troncoso, R., Ibacache, M., ... Lavandero, S. (2011). Cardiomyocyte death: Mechanisms and translational implications. *Cell Death & Disease, 2,* e244. Available from https://doi.org/10.1038/cddis.2011.130.

Damle, S. S., Moore, E. E., Babu, A. N., Meng, X., Fullerton, D. A., & Banerjee, A. (2009). Hemoglobin-based oxygen carrier induces heme oxygenase-1 in the heart and lung but not brain. *Journal of the American College of Surgeons, 208*(4), 592–598. Available from https://doi.org/10.1016/j.jamcollsurg.2009.01.015.

Deitch, E. A. (2010). Gut lymph and lymphatics: A source of factors leading to organ injury and dysfunction. *Annals of New York Academy of Sciences, 1207*(Suppl 1), E103–111. Available from https://doi.org/10.1111/j.1749-6632.2010.05713.x.

Deitch, E. A., Xu, D., & Kaise, V. L. (2006). Role of the gut in the development of injury- and shock induced SIRS and MODS: The gut–lymph hypothesis, a review. *Frontiers in Bioscience, 11,* 520–528.

Depre, C., & Vatner, S. F. (2005). Mechanisms of cell survival in myocardial hibernation. *Trends in Cardiovascular Medicine, 15*(3), 101–110. Available from https://doi.org/10.1016/j.tcm.2005.04.006.

Desmet, W., Bogaert, J., Dubois, C., Sinnaeve, P., Adriaenssens, T., Pappas, C., ... Van de Werf, F. (2011). High-dose intracoronary adenosine for myocardial salvage in patients with acute ST-segment elevation myocardial infarction. *European Heart Journal, 32*(7), 867—877. Available from https://doi.org/10.1093/eurheartj/ehq492.

Dirnagl, U., Iadecola, C., & Moskowitz, M. A. (1999). Pathobiology of ischaemic stroke: An integrated view. *Trends in Neuroscience, 22*(9), 391—397.

Dodd-o, J. M., Hristopoulos, M. L., Welsh-Servinsky, L. E., Tankersley, C. G., & Pearse, D. B. (2006). Strain-specific differences in sensitivity to ischemia-reperfusion lung injury in mice. *Journal of Applied Physiology (1985), 100*(5), 1590—1595. Available from https://doi.org/10.1152/japplphysiol.00681.2005.

Dong, Y., Undyala, V. V., Gottlieb, R. A., Mentzer, R. M., & Przyklenk, K. (2010). Review: Autophagy: Definition, molecular machinery, and potential role in myocardial ischemia-reperfusion injury. *Journal of Cardiovascular Pharmacology and Therapeutics, 15*(3), 220—230. Available from https://doi.org/10.1177/1074248410370327.

Duncker, D. J., Klassen, C. L., Ishibashi, Y., Herrlinger, S. H., Pavek, T. J., & Bache, R. J. (1996). Effect of temperature on myocardial infarction in swine. *American Journal of Physiology, 270*(4 Pt 2), H1189—H1199. Available from https://doi.org/10.1152/ajpheart.1996.270.4.H1189.

Eapen, Z. J., Tang, W. H., Felker, G. M., Hernandez, A. F., Mahaffey, K. W., Lincoff, A. M., & Roe, M. T. (2012). Defining heart failure end points in ST-segment elevation myocardial infarction trials: Integrating past experiences to chart a path forward. *Circulation: Cardiovascular Quality and Outcomes, 5*(4), 594—600. Available from https://doi.org/10.1161/circoutcomes.112.966150.

Eitel, I., de Waha, S., Wohrle, J., Fuernau, G., Lurz, P., Pauschinger, M., . . . Thiele, H. (2014). Comprehensive prognosis assessment by CMR imaging after ST-segment elevation myocardial infarction. *Journal of the American College of Cardiology, 64*(12), 1217—1226. Available from https://doi.org/10.1016/j.jacc.2014.06.1194.

Esme, H., Fidan, H., Koken, T., & Solak, O. (2006). Effect of lung ischemia—reperfusion on oxidative stress parameters of remote tissues. *European Journal of Cardio-Thoracic Surgery, 29*(3), 294—298. Available from https://doi.org/10.1016/j.ejcts.2005.12.008.

Ferdinandy, P., Hausenloy, D. J., Heusch, G., Baxter, G. F., & Schulz, R. (2014). Interaction of risk factors, comorbidities, and comedications with ischemia/reperfusion injury and cardioprotection by preconditioning, postconditioning, and remote conditioning. *Pharmacological Reviews, 66*(4), 1142—1174. Available from https://doi.org/10.1124/pr.113.008300.

Ferraro, F. J., Rush, B. F., Jr., Simonian, G. T., Bruce, C. J., Murphy, T. F., Hsieh, J. T., . . . Condon, M. (1995). A comparison of survival at different degrees of hemorrhagic shock in germ-free and germ-bearing rats. *Shock, 4*(2), 117—120.

Frangogiannis, N. G. (2008). The immune system and cardiac repair. *Pharmacological Research, 58*(2), 88—111. Available from https://doi.org/10.1016/j.phrs.2008.06.007.

Galinanes, M., & Hearse, D. J. (1990). Species differences in susceptibility to ischemic injury and responsiveness to myocardial protection. *Cardioscience, 1*(2), 127—143.

Gedik, N., Thielmann, M., Kottenberg, E., Peters, J., Jakob, H., Heusch, G., & Kleinbongard, P. (2014). No evidence for activated autophagy in left ventricular myocardium at early reperfusion with protection by remote ischemic preconditioning in patients undergoing coronary artery bypass grafting. *PLoS One, 9*(5), e96567. Available from https://doi.org/10.1371/journal.pone.0096567.

Ginks, W. R., Sybers, H. D., Maroko, P. R., Covell, J. W., Sobel, B. E., & Ross, J., Jr. (1972). Coronary artery reperfusion. II. Reduction of myocardial infarct size at 1 week after the coronary occlusion. *Journal of Clinical Investigation*, *51*(10), 2717–2723. Available from https://doi.org/10.1172/jci107091.

Granger, D. N. (1988). Role of xanthine oxidase and granulocytes in ischemia-reperfusion injury. *American Journal of Physiology*, *255*(6 Pt 2), H1269–H1275. Available from https://doi.org/10.1152/ajpheart.1988.255.6.H1269.

Granger, D. N., & Korthuis, R. J. (1995). Physiologic mechanisms of postischemic tissue injury. *Annual Review of Physiology*, *57*, 311–332. Available from https://doi.org/10.1146/annurev.ph.57.030195.001523.

Gregorini, L., Marco, J., & Heusch, G. (2012). Peri-interventional coronary vasomotion. *Journal of Molecular and Cell Cardiology*, *52*(4), 883–889. Available from https://doi.org/10.1016/j.yjmcc.2011.09.017.

Gregorini, L., Marco, J., Kozakova, M., Palombo, C., Anguissola, G. B., Marco, I., . . . Heusch, G. (1999). Alpha-adrenergic blockade improves recovery of myocardial perfusion and function after coronary stenting in patients with acute myocardial infarction. *Circulation*, *99*(4), 482–490.

Guo, Y., Flaherty, M. P., Wu, W. J., Tan, W., Zhu, X., Li, Q., & Bolli, R. (2012). Genetic background, gender, age, body temperature, and arterial blood pH have a major impact on myocardial infarct size in the mouse and need to be carefully measured and/or taken into account: Results of a comprehensive analysis of determinants of infarct size in 1,074 mice. *Basic Research in Cardiology*, *107*(5), 288. Available from https://doi.org/10.1007/s00395-012-0288-y.

Hacke, W., Donnan, G., Fieschi, C., Kaste, M., von Kummer, R., Broderick, J. P., . . . Hamilton, S. (2004). Association of outcome with early stroke treatment: Pooled analysis of ATLANTIS, ECASS, and NINDS rt-PA stroke trials. *The Lancet*, *363*(9411), 768–774. Available from https://doi.org/10.1016/s0140-6736(04)15692-4.

He, G. Z., Dong, L. G., Chen, X. F., Zhou, K. G., & Shu, H. (2011). Lymph duct ligation during ischemia/reperfusion prevents pulmonary dysfunction in a rat model with omega-3 polyunsaturated fatty acid and glutamine. *Nutrition*, *27*(5), 604–614. Available from https://doi.org/10.1016/j.nut.2010.06.003.

Hedström, E., Engblom, H., Frogner, F., Åström-Olsson, K., Öhlin, H., Jovinge, S., & Arheden, H. (2009). Infarct evolution in man studied in patients with first-time coronary occlusion in comparison to different species - implications for assessment of myocardial salvage. *Journal of Cardiovascular Magnetic Resonance*, *11*(1), 38. Available from https://doi.org/10.1186/1532-429x-11-38.

Heusch, G. (2001). Nitroglycerin and delayed preconditioning in humans: Yet another new mechanism for an old drug? *Circulation*, *103*(24), 2876–2878.

Heusch, G. (2004). Postconditioning: Old wine in a new bottle? *Journal of the American College Cardiology*, *44*(5), 1111–1112. Available from https://doi.org/10.1016/j.jacc.2004.06.013.

Heusch, G. (2008). Heart rate in the pathophysiology of coronary blood flow and myocardial ischaemia: Benefit from selective bradycardic agents. *British Journal of Pharmacology*, *153*(8), 1589–1601. Available from https://doi.org/10.1038/sj.bjp.0707673.

Heusch, G. (2013). Cardioprotection: Chances and challenges of its translation to the clinic. *The Lancet*, *381*(9861), 166–175. Available from https://doi.org/10.1016/s0140-6736(12)60916-7.

Heusch, G., Baumgart, D., Camici, P., Chilian, W., Gregorini, L., Hess, O., . . . Rimoldi, O. (2000). alpha-adrenergic coronary vasoconstriction and myocardial ischemia in humans. *Circulation*, *101*(6), 689−694.

Heusch, G., Kleinbongard, P., Bose, D., Levkau, B., Haude, M., Schulz, R., & Erbel, R. (2009). Coronary microembolization: From bedside to bench and back to bedside. *Circulation*, *120*(18), 1822−1836. Available from https://doi.org/10.1161/circulationaha.109.888784.

Heusch, G., Kleinbongard, P., & Skyschally, A. (2013). Myocardial infarction and coronary microvascular obstruction: An intimate, but complicated relationship. *Basic Research in Cardiology*, *108*(6), 380. Available from https://doi.org/10.1007/s00395-013-0380-y.

Heusch, G., Kleinbongard, P., Skyschally, A., Levkau, B., Schulz, R., & Erbel, R. (2012). The coronary circulation in cardioprotection: More than just one confounder†. *Cardiovascular Research*, *94*(2), 237−245. Available from https://doi.org/10.1093/cvr/cvr271.

Heusch, G., Libby, P., Gersh, B., Yellon, D., Bohm, M., Lopaschuk, G., & Opie, L. (2014). Cardiovascular remodelling in coronary artery disease and heart failure. *The Lancet*, *383*(9932), 1933−1943. Available from https://doi.org/10.1016/s0140-6736(14)60107-0.

Heusch, G., & Schulz, R. (2002). Hibernating myocardium: New answers, still more questions!. *Circulation Research*, *91*(10), 863−865.

Heusch, G., Schulz, R., & Rahimtoola, S. H. (2005). Myocardial hibernation: A delicate balance. *American Journal of Physiology—Heart and Circulatory Physiology*, *288*(3), H984−H999. Available from https://doi.org/10.1152/ajpheart.01109.2004.

Hirsch, J., Niemann, C. U., Hansen, K. C., Choi, S., Su, X., Frank, J. A., & Matthay, M. A. (2008). Alterations in the proteome of pulmonary alveolar type II cells in the rat after hepatic ischemia-reperfusion. *Critical Care Medicine*, *36*(6), 1846−1854. Available from https://doi.org/10.1097/CCM.0b013e31816f49cb.

Horneffer, P. J., Healy, B., Gott, V. L., & Gardner, T. J. (1987). The rapid evolution of a myocardial infarction in an end-artery coronary preparation. *Circulation*, *76*(5 Pt 2), V39−V42.

Humphreys, M. R., Castle, E. P., Lohse, C. M., Sebo, T. J., Leslie, K. O., & Andrews, P. E. (2009). Renal ischemia time in laparoscopic surgery: An experimental study in a porcine model. *International Journal of Urology*, *16*(1), 105−109. Available from https://doi.org/10.1111/j.1442-2042.2008.02173.x.

Ibáñez, B., Heusch, G., Ovize, M., & Van de Werf, F. (2015). Evolving therapies for myocardial ischemia/reperfusion injury. *Journal of the American College of Cardiology*, *65*(14), 1454−1471. Available from https://doi.org/10.1016/j.jacc.2015.02.032.

Ibanez, B., Macaya, C., Sanchez-Brunete, V., Pizarro, G., Fernandez-Friera, L., Mateos, A., . . . Fuster, V. (2013). Effect of early metoprolol on infarct size in ST-segment-elevation myocardial infarction patients undergoing primary percutaneous coronary intervention: The Effect of Metoprolol in Cardioprotection During an Acute Myocardial Infarction (METOCARD-CNIC) trial. *Circulation*, *128*(14), 1495−1503. Available from https://doi.org/10.1161/circulationaha.113.003653.

Ikeda, H., Suzuki, Y., Suzuki, M., Koike, M., Tamura, J., Tong, J., . . . Itoh, G. (1998). Apoptosis is a major mode of cell death caused by ischaemia and ischaemia/reperfusion injury to the rat intestinal epithelium. *Gut*, *42*(4), 530−537.

Inserte, J., Hernando, V., & Garcia-Dorado, D. (2012). Contribution of calpains to myocardial ischaemia/reperfusion injury. *Cardiovascular Research*, *96*(1), 23−31. Available from https://doi.org/10.1093/cvr/cvs232.

Jang, H. R., Ko, G. J., Wasowska, B. A., & Rabb, H. (2009). The interaction between ischemia-reperfusion and immune responses in the kidney. *Journal of Molecular Medicine (Berlin, Germany)*, *87*(9), 859−864. Available from https://doi.org/10.1007/s00109-009-0491-y.

Jennings, R. B., Sommers, H. M., Smyth, G. A., Flack, H. A., & Linn, H. (1960). Myocardial necrosis induced by temporary occlusion of a coronary artery in the dog. *Archives of Pathology*, *70*, 68−78.

Julia, P. L., Kofsky, E. R., Buckberg, G. D., Young, H. H., & Bugyi, H. I. (1990). Studies of myocardial protection in the immature heart. I. Enhanced tolerance of immature versus adult myocardium to global ischemia with reference to metabolic differences. *The Journal of Thoracic and Cardiovascular Surgery*, *100*(6), 879−887.

Kalogeris, T., Baines, C. P., Krenz, M., & Korthuis, R. J. (2012). Cell biology of ischemia/reperfusion injury. *International review of cell and molecular biology*, *298*, 229−317.

Kassahun, W. T., Schulz, T., Richter, O., & Hauss, J. (2008). Unchanged high mortality rates from acute occlusive intestinal ischemia: Six year review. *Langenbeck's Archives of Surgery*, *393*(2), 163−171. Available from https://doi.org/10.1007/s00423-007-0263-5.

Kinross, J., Warren, O., Basson, S., Holmes, E., Silk, D., Darzi, A., & Nicholson, J. K. (2009). Intestinal ischemia/reperfusion injury: Defining the role of the gut microbiome. *Biomarkers in Medicine*, *3*(2), 175−192. Available from https://doi.org/10.2217/bmm.09.11.

Kinross, J. M., Darzi, A. W., & Nicholson, J. K. (2011). Gut microbiome-host interactions in health and disease. *Genome Medicine*, *3*(3), 14. Available from https://doi.org/10.1186/gm228.

Kleinbongard, P., Baars, T., Mohlenkamp, S., Kahlert, P., Erbel, R., & Heusch, G. (2013). Aspirate from human stented native coronary arteries vs. saphenous vein grafts: More endothelin but less particulate debris. *American Journal of Physiology—Heart and Circulatory Physiology*, *305*(8), H1222−H1229. Available from https://doi.org/10.1152/ajpheart.00358.2013.

Kleinbongard, P., Bose, D., Baars, T., Mohlenkamp, S., Konorza, T., Schoner, S., ... Heusch, G. (2011). Vasoconstrictor potential of coronary aspirate from patients undergoing stenting of saphenous vein aortocoronary bypass grafts and its pharmacological attenuation. *Circulation Research*, *108*(3), 344−352. Available from https://doi.org/10.1161/circresaha.110.235713.

Kleinbongard, P., Schulz, R., & Heusch, G. (2011). TNFα in myocardial ischemia/reperfusion, remodeling and heart failure. *Heart Failure Reviews*, *16*(1), 49−69. Available from https://doi.org/10.1007/s10741-010-9180-8.

Kloner, R. A. (2013). Current state of clinical translation of cardioprotective agents for acute myocardial infarction. *Circulation Research*, *113*(4), 451−463. Available from https://doi.org/10.1161/circresaha.112.300627.

Kloner, R. A., Ellis, S. G., Lange, R., & Braunwald, E. (1983). Studies of experimental coronary artery reperfusion. Effects on infarct size, myocardial function, biochemistry, ultrastructure and microvascular damage. *Circulation*, *68*(2 Pt 2), I8−I15.

Kloner, R. A., Ganote, C. E., & Jennings, R. B. (1974). The "no-reflow" phenomenon after temporary coronary occlusion in the dog. *Journal of Clinical Investigation, 54*(6), 1496–1508. Available from https://doi.org/10.1172/jci107898.

Kloner, R. A., Rude, R. E., Carlson, N., Maroko, P. R., DeBoer, L. W., & Braunwald, E. (1980). Ultrastructural evidence of microvascular damage and myocardial cell injury after coronary artery occlusion: Which comes first? *Circulation, 62*(5), 945–952.

Kristian, T. (2004). Metabolic stages, mitochondria and calcium in hypoxic/ischemic brain damage. *Cell Calcium, 36*(3-4), 221–233. Available from https://doi.org/10.1016/j.ceca.2004.02.016.

Kroemer, G., & Reed, J. C. (2000). Mitochondrial control of cell death. *Nature Medicine, 6*, 513. Available from https://doi.org/10.1038/74994.

Ladilov, Y. V., Siegmund, B., & Piper, H. M. (1995). Protection of reoxygenated cardiomyocytes against hypercontracture by inhibition of Na^+/H^+ exchange. *American Journal of Physiology, 268*(4 Pt 2), H1531–H1539. Available from https://doi.org/10.1152/ajpheart.1995.268.4.H1531.

Lam, V., Su, J., Koprowski, S., Hsu, A., Tweddell, J. S., Rafiee, P., ... Baker, J. E. (2012). Intestinal microbiota determine severity of myocardial infarction in rats. *FASEB Journal, 26*(4), 1727–1735. Available from https://doi.org/10.1096/fj.11-197921.

LaPointe, N. M., Al-Khatib, S. M., Piccini, J. P., Atwater, B. D., Honeycutt, E., Thomas, K., ... Peterson, E. D. (2011). Extent of and reasons for nonuse of implantable cardioverter defibrillator devices in clinical practice among eligible patients with left ventricular systolic dysfunction. *Circulation: Cardiovascular Quality and Outcomes, 4*(2), 146–151. Available from https://doi.org/10.1161/circoutcomes.110.958603.

Larose, E., Rodes-Cabau, J., Pibarot, P., Rinfret, S., Proulx, G., Nguyen, C. M., ... Bertrand, O. F. (2010). Predicting late myocardial recovery and outcomes in the early hours of ST-segment elevation myocardial infarction traditional measures compared with microvascular obstruction, salvaged myocardium, and necrosis characteristics by cardiovascular magnetic resonance. *Journal of the Americal College of Cardiology, 55*(22), 2459–2469. Available from https://doi.org/10.1016/j.jacc.2010.02.033.

LATE Study Group (1993). Late Assessment of Thrombolytic Efficacy (LATE) study with alteplase 6–24 hours after onset of acute myocardial infarction. *The Lancet, 342*(8874), 759–766. Available from https://www.ncbi.nlm.nih.gov/pubmed/8103874

Lee, J. M., Grabb, M. C., Zipfel, G. J., & Choi, D. W. (2000). Brain tissue responses to ischemia. *Journal of Clinical Investigation, 106*(6), 723–731. Available from https://doi.org/10.1172/jci11003.

Liu, L., & Kubes, P. (2003). Molecular mechanisms of leukocyte recruitment: Organ-specific mechanisms of action. *Thrombosis and Haemostasis, 89*(2), 213–220.

Manciet, L. H., Poole, D. C., McDonagh, P. F., Copeland, J. G., & Mathieu-Costello, O. (1994). Microvascular compression during myocardial ischemia: Mechanistic basis for no-reflow phenomenon. *American Journal of Physiology, 266*(4 Pt 2), H1541–H1550. Available from https://doi.org/10.1152/ajpheart.1994.266.4.H1541.

Manintveld, O. C., Te Lintel Hekkert, M., van den Bos, E. J., Suurenbroek, G. M., Dekkers, D. H., Verdouw, P. D., ... Duncker, D. J. (2007). Cardiac effects of postconditioning depend critically on the duration of index ischemia. *American Journal of*

Physiology—Heart and Circulatory Physiology, 292(3), H1551–H1560. Available from https://doi.org/10.1152/ajpheart.00151.2006.

Maroko, P. R., Libby, P., Ginks, W. R., Bloor, C. M., Shell, W. E., Sobel, B. E., & Ross, J., Jr. (1972). Coronary artery reperfusion. I. Early effects on local myocardial function and the extent of myocardial necrosis. *Journal of Clinical Investigation, 51*(10), 2710–2716. Available from https://doi.org/10.1172/jci107090.

McDougal, W. S. (1988). Renal perfusion/reperfusion injuries. *Journal of Urology, 140*(6), 1325–1330.

Menees, D. S., Peterson, E. D., Wang, Y., Curtis, J. P., Messenger, J. C., Rumsfeld, J. S., & Gurm, H. S. (2013). Door-to-balloon time and mortality among patients undergoing primary PCI. *New England Journal of Medicine, 369*(10), 901–909. Available from https://doi.org/10.1056/NEJMoa1208200.

Mewton, N., Thibault, H., Roubille, F., Lairez, O., Rioufol, G., Sportouch, C., ... Ovize, M. (2013). Postconditioning attenuates no-reflow in STEMI patients. *Basic Research in Cardiology, 108*(6), 383. Available from https://doi.org/10.1007/s00395-013-0383-8.

Michael, L. H., Entman, M. L., Hartley, C. J., Youker, K. A., Zhu, J., Hall, S. R., ... Ballantyne, C. M. (1995). Myocardial ischemia and reperfusion: A murine model. *American Journal of Physiology, 269*(6 Pt 2), H2147–H2154. Available from https://doi.org/10.1152/ajpheart.1995.269.6.H2147.

Miura, T., Yellon, D. M., Hearse, D. J., & Downey, J. M. (1987). Determinants of infarct size during permanent occlusion of a coronary artery in the closed chest dog. *Journal of the American College of Cardiology, 9*(3), 647–654.

Murry, C. E., Jennings, R. B., & Reimer, K. A. (1986). Preconditioning with ischemia: A delay of lethal cell injury in ischemic myocardium. *Circulation, 74*(5), 1124–1136.

Niccoli, G., Burzotta, F., Galiuto, L., & Crea, F. (2009). Myocardial no-reflow in humans. *Journal of the American College of Cardiology, 54*(4), 281–292. Available from https://doi.org/10.1016/j.jacc.2009.03.054.

Oerlemans, M. I., Koudstaal, S., Chamuleau, S. A., de Kleijn, D. P., Doevendans, P. A., & Sluijter, J. P. (2013). Targeting cell death in the reperfused heart: Pharmacological approaches for cardioprotection. *International Journal of Cardiology, 165*(3), 410–422. Available from https://doi.org/10.1016/j.ijcard.2012.03.055.

O'Gara, P. T., Kushner, F. G., Ascheim, D. D., Casey, D. E., Jr., Chung, M. K., de Lemos, J. A., ... Zhao, D. X. (2013). 2013 ACCF/AHA guideline for the management of ST-elevation myocardial infarction: Executive summary: A report of the American College of Cardiology Foundation/American Heart Association Task Force on Practice Guidelines: Developed in collaboration with the American College of Emergency Physicians and Society for Cardiovascular Angiography and Interventions. *Catheterization and Cardiovascular Interventions, 82*(1), E1–E27. Available from https://doi.org/10.1002/ccd.24776.

Okamoto, F., Allen, B. S., Buckberg, G. D., Bugyi, H., & Leaf, J. (1986). Reperfusion conditions: Importance of ensuring gentle versus sudden reperfusion during relief of coronary occlusion. *The Journal of Thoracic and Cardiovascular Surgery, 92*(3 Pt 2), 613–620.

Ordy, J. M., Wengenack, T. M., Bialobok, P., Coleman, P. D., Rodier, P., Baggs, R. B., ... Kates, B. (1993). Selective vulnerability and early progression of hippocampal CA1 pyramidal cell degeneration and GFAP-positive astrocyte reactivity in the rat four-vessel occlusion model of transient global ischemia. *Experimental Neurology, 119*(1), 128—139. Available from https://doi.org/10.1006/exnr.1993.1014.

Orogo, A. M., & Gustafsson, A. B. (2013). Cell death in the myocardium: My heart won't go on. *IUBMB Life, 65*(8), 651—656. Available from https://doi.org/10.1002/iub.1180.

Park, P. O., & Haglund, U. (1992). Regeneration of small bowel mucosa after intestinal ischemia. *Critical Care in Medicine, 20*(1), 135—139.

Pich, S., Klein, H. H., Lindert, S., Nebendahl, K., & Kreuzer, H. (1988). Cell death in ischemic, reperfused porcine hearts: A histochemical and functional study. *Basic Research in Cardiology, 83*(5), 550—559. Available from https://doi.org/10.1007/BF01906684.

Piot, C., Croisille, P., Staat, P., Thibault, H., Rioufol, G., Mewton, N., ... Ovize, M. (2008). Effect of cyclosporine on reperfusion injury in acute myocardial infarction. *New England Journal of Medicine, 359*(5), 473—481. Available from https://doi.org/10.1056/NEJMoa071142.

Piper, H. M., Abdallah, Y., & Schafer, C. (2004). The first minutes of reperfusion: A window of opportunity for cardioprotection. *Cardiovascular Research, 61*(3), 365—371. Available from https://doi.org/10.1016/j.cardiores.2003.12.012.

Piper, H. M., Meuter, K., & Schafer, C. (2003). Cellular mechanisms of ischemia-reperfusion injury. *The Annals of Thoracic Surgery, 75*(2), S644—S648.

Premen, A. J., Banchs, V., Womack, W. A., Kvietys, P. R., & Granger, D. N. (1987). Importance of collateral circulation in the vascularly occluded feline intestine. *Gastroenterology, 92*(5 Pt 1), 1215—1219.

Przyklenk, K., Reddy Undyala, V. V., Wider, J., Sala-Mercado, J. A., Gottlieb, R. A., & Mentzer, J. R. M. (2011). Acute induction of autophagy as a novel strategy for cardio-protection. *Autophagy, 7*(4), 432—433. Available from https://doi.org/10.4161/auto.7.4.14395.

Quartara, L., & Maggi, C. A. (1998). The tachykinin NK1 receptor. Part II: Distribution and pathophysiological roles. *Neuropeptides, 32*(1), 1—49.

Redel, A., Jazbutyte, V., Smul, T. M., Lange, M., Eckle, T., Eltzschig, H., ... Kehl, F. (2008). Impact of ischemia and reperfusion times on myocardial infarct size in mice in vivo. *Experimental Biology and Medicine, 233*(1), 84—93. Available from https://doi.org/10.3181/0612-RM-308.

Reffelmann, T., & Kloner, R. A. (2002). Microvascular reperfusion injury: Rapid expansion of anatomic no reflow during reperfusion in the rabbit. *American Journal of Physiology—Heart and Circulatory Physiology, 283*(3), H1099—H1107. Available from https://doi.org/10.1152/ajpheart.00270.2002.

Reimer, K. A., & Jennings, R. B. (1979). The "wavefront phenomenon" of myocardial ischemic cell death. II. Transmural progression of necrosis within the framework of ischemic bed size (myocardium at risk) and collateral flow. *Laboratory Investigation, 40*(6), 633—644.

Reimer, K. A., Jennings, R. B., & Tatum, A. H. (1983). Pathobiology of acute myocardial ischemia: Metabolic, functional and ultrastructural studies. *American Journal of Cardiology, 52*(2), 72A—81A.

Reimer, K. A., Lowe, J. E., Rasmussen, M. M., & Jennings, R. B. (1977). The wavefront phenomenon of ischemic cell death. 1. Myocardial infarct size vs duration of coronary occlusion in dogs. *Circulation, 56*(5), 786—794.

Rezkalla, S. H., & Kloner, R. A. (2004). Ischemic preconditioning and preinfarction angina in the clinical arena. *Nature Clinical Practice Cardiovascular Medicine*, *1*, 96. Available from https://doi.org/10.1038/ncpcardio0047.

Roberts, C. S., Schoen, F. J., & Kloner, R. A. (1983). Effect of coronary reperfusion on myocardial hemorrhage and infarct healing. *American Journal of Cardiology*, *52*(5), 610–614.

Rochitte, C. E., Lima, J. A., Bluemke, D. A., Reeder, S. B., McVeigh, E. R., Furuta, T., ... Melin, J. A. (1998). Magnitude and time course of microvascular obstruction and tissue injury after acute myocardial infarction. *Circulation*, *98*(10), 1006–1014.

Roubille, F., Lairez, O., Mewton, N., Rioufol, G., Ranc, S., Sanchez, I., ... Ovize, M. (2012). Cardioprotection by clopidogrel in acute ST-elevated myocardial infarction patients: A retrospective analysis. *Basic Research in Cardiology*, *107*(4), 275. Available from https://doi.org/10.1007/s00395-012-0275-3.

Sala-Mercado, J. A., Wider, J., Undyala, V. V., Jahania, S., Yoo, W., Mentzer, R. M., Jr., ... Przyklenk, K. (2010). Profound cardioprotection with chloramphenicol succinate in the swine model of myocardial ischemia-reperfusion injury. *Circulation*, *122*(11 Suppl), S179–S184. Available from https://doi.org/10.1161/circulationaha.109.928242.

Santora, R. J., Lie, M. L., Grigoryev, D. N., Nasir, O., Moore, F. A., & Hassoun, H. T. (2010). Therapeutic distant organ effects of regional hypothermia during mesenteric ischemia-reperfusion injury. *Journal of Vascular Surgery*, *52*(4), 1003–1014. Available from https://doi.org/10.1016/j.jvs.2010.05.088.

Sapega, A. A., Heppenstall, R. B., Chance, B., Park, Y. S., & Sokolow, D. (1985). Optimizing tourniquet application and release times in extremity surgery. A biochemical and ultrastructural study. *Journal of Bone and Joint Surgery American*, *67*(2), 303–314.

Schaper, W., Gorge, G., Winkler, B., & Schaper, J. (1988). The collateral circulation of the heart. *Progress in Cardiovascular Disease*, *31*(1), 57–77.

Schluter, K. D., Jakob, G., Ruiz-Meana, M., Garcia-Dorado, D., & Piper, H. M. (1996). Protection of reoxygenated cardiomyocytes against osmotic fragility by nitric oxide donors. *American Journal of Physiology*, *271*(2 Pt 2), H428–H434. Available from https://doi.org/10.1152/ajpheart.1996.271.2.H428.

Schömig, A., Mehilli, J., Antoniucci, D., Ndrepepa, G., Markwardt, C., Di Pede, F., ... Kastrati, A. (2005). Mechanical reperfusion in patients with acute myocardial infarction presenting more than 12 hours from symptom onset: A randomized controlled trial. *JAMA*, *293*(23), 2865–2872. Available from https://doi.org/10.1001/jama.293.23.2865.

Seiler, C., Stoller, M., Pitt, B., & Meier, P. (2013). The human coronary collateral circulation: Development and clinical importance. *European Heart Journal*, *34*(34), 2674–2682. Available from https://doi.org/10.1093/eurheartj/eht195.

Singh, K. K., Yanagawa, B., Quan, A., Wang, R., Garg, A., Khan, R., ... Verma, S. (2014). Autophagy gene fingerprint in human ischemia and reperfusion. *The Journal of Thoracic and Cardiovascular Surgery*, *147*(3), 1065–1072.e1061. Available from https://doi.org/10.1016/j.jtcvs.2013.04.042.

Skyschally, A., Schulz, R., & Heusch, G. (2008). Pathophysiology of myocardial infarction: Protection by ischemic pre- and postconditioning. *Herz*, *33*(2), 88–100. Available from https://doi.org/10.1007/s00059-008-3101-9.

Skyschally, A., Walter, B., & Heusch, G. (2013). Coronary microembolization during early reperfusion: Infarct extension, but protection by ischaemic postconditioning.

European Heart Journal, 34(42), 3314—3321. Available from https://doi.org/10.1093/eurheartj/ehs434.

Slezak, J., Tribulova, N., Okruhlicova, L., Dhingra, R., Bajaj, A., Freed, D., & Singal, P. (2009). Hibernating myocardium: Pathophysiology, diagnosis, and treatment. *Canadian Journal of Physiology and Pharmacology, 87*(4), 252—265. Available from https://doi.org/10.1139/y09-011.

Smith, V. A., & Johnson, T. (2010). Evaluation of an animal product-free variant of MegaCell MEM as a storage medium for corneas destined for transplantation. *Ophthalmic Research, 43*(1), 33—42. Available from https://doi.org/10.1159/000246576.

Sorkine, P., Szold, O., Halpern, P., Gutman, M., Greemland, M., Rudick, V., & Goldman, G. (1997). Gut decontamination reduces bowel ischemia-induced lung injury in rats. *Chest, 112*(2), 491—495.

Souza, D. G., Mendonca, V. A., de, A. C. M. S., Poole, S., & Teixeira, M. M. (2002). Role of tachykinin NK receptors on the local and remote injuries following ischaemia and reperfusion of the superior mesenteric artery in the rat. *British Journal of Pharmacology, 135*(2), 303—312. Available from https://doi.org/10.1038/sj.bjp.0704464.

Souza, D. G., Vieira, A. T., Soares, A. C., Pinho, V., Nicoli, J. R., Vieira, L. Q., & Teixeira, M. M. (2004). The essential role of the intestinal microbiota in facilitating acute inflammatory responses. *Journal of Immunology, 173*(6), 4137—4146.

Staat, P., Rioufol, G., Piot, C., Cottin, Y., Cung, T. T., L'Huillier, I., ... Ovize, M. (2005). Postconditioning the human heart. *Circulation, 112*(14), 2143—2148. Available from https://doi.org/10.1161/circulationaha.105.558122.

Stallion, A., Kou, T. D., Latifi, S. Q., Miller, K. A., Dahms, B. B., Dudgeon, D. L., & Levine, A. D. (2005). Ischemia/reperfusion: A clinically relevant model of intestinal injury yielding systemic inflammation. *Journal of Pediatric Surgery, 40*(3), 470—477. Available from https://doi.org/10.1016/j.jpedsurg.2004.11.045.

Tani, M., & Neely, J. R. (1989). Role of intracellular Na^+ in Ca^{2+} overload and depressed recovery of ventricular function of reperfused ischemic rat hearts. Possible involvement of $H + -Na^+$ and Na^+-Ca^{2+} exchange. *Circulation Research, 65*(4), 1045—1056.

Thibault, H., Gomez, L., Donal, E., Pontier, G., Scherrer-Crosbie, M., Ovize, M., & Derumeaux, G. (2007). Acute myocardial infarction in mice: Assessment of transmurality by strain rate imaging. *American Journal of Physiology—Heart and Circulatory Physiology, 293*(1), H496—H502. Available from https://doi.org/10.1152/ajpheart.00087.2007.

Vaghasiya, J. D., Sheth, N. R., Bhalodia, Y. S., & Jivani, N. P. (2010). Exaggerated liver injury induced by renal ischemia reperfusion in diabetes: Effect of exenatide. *Saudi Journal of Gastroenterology, 16*(3), 174—180. Available from https://doi.org/10.4103/1319-3767.65187.

Van de Werf, F. (2014). The history of coronary reperfusion. *European Heart Journal, 35*(37), 2510—2515. Available from https://doi.org/10.1093/eurheartj/ehu268.

Vega, V. L., Mardones, L., Maldonado, M., Nicovani, S., Manriquez, V., Roa, J., & Ward, P. H. (2000). Xanthine oxidase released from reperfused hind limbs mediate kupffer cell activation, neutrophil sequestration, and hepatic oxidative stress in rats subjected to tourniquet shock. *Shock, 14*(5), 565—571.

Wagers, A. J., & Conboy, I. M. (2005). Cellular and molecular signatures of muscle regeneration: Current concepts and controversies in adult myogenesis. *Cell, 122*(5), 659−667. Available from https://doi.org/10.1016/j.cell.2005.08.021.

Wang, B., Huang, Q., Zhang, W., Li, N., & Li, J. (2011). *Lactobacillus plantarum* prevents bacterial translocation in rats following ischemia and reperfusion injury. *Digestive Diseases and Sciences, 56*(11), 3187−3194. Available from https://doi.org/10.1007/s10620-011-1747-2.

Weber, C., & Noels, H. (2011). Atherosclerosis: Current pathogenesis and therapeutic options. *Nature Medicine, 17*, 1410. Available from https://doi.org/10.1038/nm.2538.

Weiss, J. N., Korge, P., Honda, H. M., & Ping, P. (2003). Role of the mitochondrial permeability transition in myocardial disease. *Circulation Research, 93*(4), 292.

Willems, I. E., Havenith, M. G., De Mey, J. G., & Daemen, M. J. (1994). The alpha-smooth muscle actin-positive cells in healing human myocardial scars. *The American Journal of Pathology, 145*(4), 868−875.

Winquist, R. J., & Kerr, S. (1997). Cerebral ischemia-reperfusion injury and adhesion. *Neurology, 49*(5 Suppl 4), S23−S26.

Wittnich, C. (1992). Age-related differences in myocardial metabolism affects response to ischemia. Age in heart tolerance to ischemia. *The American Journal of Cardiovascular Pathology, 4*(2), 175−180.

Yancy, C. W., Jessup, M., Bozkurt, B., Butler, J., Casey, D. E., Jr., Drazner, M. H., ... Wilkoff, B. L. (2013). 2013 ACCF/AHA guideline for the management of heart failure: A report of the American College of Cardiology Foundation/American Heart Association Task Force on Practice Guidelines. *Journal of the American College of Cardiology, 62*(16), e147−e239. Available from https://doi.org/10.1016/j.jacc.2013.05.019.

Yang, X.-M., Liu, Y., Liu, Y., Tandon, N., Kambayashi, J., Downey, J. M., & Cohen, M. V. (2010). Attenuation of infarction in cynomolgus monkeys: Preconditioning and postconditioning. *Basic Research in Cardiology, 105*(1), 119−128. Available from https://doi.org/10.1007/s00395-009-0050-2.

Yellon, D. M., & Hausenloy, D. J. (2007). Myocardial reperfusion injury. *The New England Journal of Medicine, 357*(11), 1121−1135. Available from https://doi.org/10.1056/NEJMra071667.

Yilmaz, G., & Granger, D. N. (2008). Cell adhesion molecules and ischemic stroke. *Neurological Research, 30*(8), 783−793. Available from https://doi.org/10.1179/174313208x341085.

Yoshiya, K., Lapchak, P. H., Thai, T. H., Kannan, L., Rani, P., Dalle Lucca, J. J., & Tsokos, G. C. (2011). Depletion of gut commensal bacteria attenuates intestinal ischemia/reperfusion injury. *American Journal of Physiology—Gastrointestinal and Liver Physiology, 301*(6), G1020−G1030. Available from https://doi.org/10.1152/ajpgi.00239.2011.

Zhang, J. H., & Xu, M. (2000). DNA fragmentation in apoptosis. *Cell Research, 10*, 205. Available from https://doi.org/10.1038/sj.cr.7290049.

Zhao, Z. Q., Corvera, J. S., Halkos, M. E., Kerendi, F., Wang, N. P., Guyton, R. A., & Vinten-Johansen, J. (2003). Inhibition of myocardial injury by ischemic postconditioning during reperfusion: Comparison with ischemic preconditioning. *American Journal of Physiology—Heart and Circulatory Physiology, 285*(2), H579−H588. Available from https://doi.org/10.1152/ajpheart.01064.2002.

Zhao, Z. Q., Nakamura, M., Wang, N. P., Velez, D. A., Hewan-Lowe, K. O., Guyton, R. A., & Vinten-Johansen, J. (2000). Dynamic progression of contractile and endothelial dysfunction and infarct extension in the late phase of reperfusion. *Journal of Surgical Research*, *94*(2), 133—144. Available from https://doi.org/10.1006/jsre.2000.6029.

Zhou, W., & Yuan, J. (2014). SnapShot: Necroptosis. *Cell*, *158*(2), 464—464.e461. Available from https://doi.org/10.1016/j.cell.2014.06.041.

Further reading

Suárez-Barrientos, A., López-Romero, P., Vivas, D., Castro-Ferreira, F., Núñez-Gil, I., Franco, E., ... Ibanez, B. (2011). Circadian variations of infarct size in acute myocardial infarction. *Heart*, *97*(12), 970—976. Available from https://doi.org/10.1136/hrt.2010.212621.

CHAPTER 2

Myocardial ischemia

Abstract

Myocardial ischemia occurs when blood flow to the heart is reduced, preventing it from receiving enough oxygen. The reduced blood flow is usually the result of a partial or complete blockage of the heart's arteries (coronary arteries). Myocardial ischemia can damage heart muscle, reducing its ability to pump efficiently. A sudden, severe blockage of a coronary artery can lead to a heart attack. Myocardial ischemia might also cause serious abnormal heart rhythms. Treatment for myocardial ischemia involves improving blood flow to the heart muscle. Treatment may include medications, a procedure to open blocked arteries, or bypass surgery.

Keywords: Myocardial ischemia; myocardial infarction; cardiac surgery

Contents

> *No understanding of the circulatory reactions of the body is possible unless we start first with the fundamental properties of the heart muscle itself, and then find out how these are modified, protected and controlled under the influence of the mechanisms — nervous, chemical, and mechanical — which under normal conditions play on the heart and blood vessels.*
>
> ***Ernest H. Starling (1919), London***

Myocardial damage due to ischemia results from an inadequate supply of oxygen and nutrients to the myocardium. It is a pathological scenarios concerning the myocardial perfusion, during cardiac surgery (Beating Heart Surgery or Off Pump Coronary Artery Bypass Surgery), and heart transplantation (Schipper et al., 2016).

Since the myocardium is highly dependent on proper coronary perfusion, the sudden interruption of the latter rapidly leads to anaerobic metabolism, depletion of high-energy phosphates, and the onset of

Pathophysiology of Ischemia Reperfusion Injury and Use of Fingolimod in Cardioprotection
DOI: https://doi.org/10.1016/B978-0-12-818023-5.00002-9
41

anaerobic glycolysis. A quick and early restoration of blood flow in ischemic tissue is, therefore, essential in all patients that present different conditions associated with local or global interruption of blood flow, to secure recovery of cells if the damage from these immediately is reversible; reperfusion, despite the undeniable utility, is associated with multiple adverse changes at the cellular level. Consequently, reperfused tissue may have myocardial damage in addition to injury caused by myocardial ischemia. Reperfusion itself can cause damage to the cardiomyocytes, increase the oxidative phenomena, and induce apoptosis; this iatrogenic damage can constitute up to 50% of the total area of necrotic myocardium (Heusch, 2013; Yellon & Hausenloy, 2007).

Therefore considering the counterproductive implications of myocardial reperfusion injury, and considering that a quick reperfusion is the best treatment of ischemia, there is absolute need for research to discover new pharmacological approaches aimed at reduction in the incidence of heart failure, preservation of left ventricular function, and to prevent cardiac remodeling to improve the survival of patients who suffered with a myocardial ischemia.

Circulatory arrest

Circulatory arrest is a state of sudden cessation of heartbeat and cardiac function, resulting in the loss of effective circulation, confirmed by no pulse, with immediate termination of the blood flow, leading to a rapid depletion of oxygen, with depression of brain function and then loss of consciousness, sleep apnea, or agonal breathing in out-hospital patients or in patients already hospitalized (Cummins et al., 1997). Cardiac arrest have be further differentiated into two subdivisions, nonfatal and fatal cardiac arrest, by responsiveness or not, to maneuvres of basic life support or advanced cardiovascular life support. The incidence of cardiac arrest in the hospital varies according to the type of construction, type of hospital patients, and the presence of a primary care team, and oscillates between 1% and 7% (Chan, Jain, Nallmothu, Berg, & Sasson, 2010; Sandroni et al., 2004) with survival that is around 20% (Bradley et al., 2012; Chen et al., 2013;Gwinnutt, Columb, & Harris, 2000; Nadkarni et al., 2006). Mortality at a distance varies according to age, pathology, and neurological status of patients after discharge from the hospital, and is higher in adults than in the pediatric population (27% vs 18%) (Nadkarni et al., 2006).

The most frequent cases of cardiac arrest (Wallmuller et al., 2012) are due to cardiac causes (63%), of which 35% are due to myocardial infarction, pulmonary causes (15%), and other causes (22%). Cardiac causes include myocardial infarction, myocardial ischemic attacks, valvular heart disease, congenital heart disease, myocarditis, atrial myxoma and cardiac temponade.

Extra cardiac causes include severe acute respiratory failure, severe hypoxemia, severe respiratory insufficiency, pulmonary edema, pulmonary embolism, dissection or rupture of the aorta, cerebral hemorrhage, marked bleeding, sepsis, serious electrolyte disturbances (hypo/hyperkalemia), and poisonings or side effects of medications or anesthetics.

Concerning cardiac causes, alterations in electrical impulses or mechanical reasons to cause ineffectiveness in cardiac function. The most common electric mechanism is represented by ventricular fibrillation, followed by asystole, pulseless electrical activity (electromechanical dissociation,) in which the ECG detects electrical activity, but there is no effective cardiac output, or pulseless ventricular tachycardia. The mechanical mechanisms include the rupture of the ventricle, pericardial tamponade, mechanical obstruction of a coronary artery and the acute rupture of a large caliber vessel. The immediate consequence is the absence of systemic perfusion. Acute myocardial infarction (Thygesen et al., 2012) is characterized by an evident myocardial necrosis that presents clinically as:

1. Patient death preceded by a typical symptom of myocardial ischemia and electrical signs of ischemia or left bundle branch block (LBBB) not present in previous ECG, before a possible check of positivity of biochemical markers of myocardial damage.
2. Raising or reduction of fingolimod markers higher than the 99th percentile of reference values, accompanied by at least one of the following:
   a. Symptoms related to myocardial ischemia.
   b. Appearance of Q waves on ECG.
   c. ECG appearance of T wave abnormalities or myocardial infarction or LBBB.
   d. The appearance of myocardial substance loss or alterations of motility of the ventricular walls highlighted with ultrasound or other myocardial imaging tests.
   e. Coronary thrombosis or any possible stent, highlighted with coronary angiography or an autopsy.

3. Following the intervention of coronary angioplasty or stent placement if:
   a. Increase troponin C ($>5 \times 99$ percentile values) in patients with normal baseline or 2% increased of baseline values for patients with baseline values increased and simultaneous presence of symptoms related to myocardial ischemia, appearance of Q waves on the ECG, appearance of alteration of the T wave, myocardial infarction, LBBB, myocardial, ventricular dyskinesias, substance loss, evidence of coronary thrombosis and eventual stent, or in the case of specific complications during the procedure.
4. Following coronary artery bypass surgery:
   a. Increase of cTn (10×99 percentile $>$ reference values) in patients with normal basal values and simultaneous presence of presence of symptoms related to myocardial ischemia, appearance of Q waves on the ECG, or appearance of alteration of the T wave, myocardial infarction, or LBBB, or loss of myocardial ventricular dyskinesias, or substance loss or evidence of coronary angiography and possible highlighted by stent or autopsy.

In the field of cardiac surgery, in adult patients, the incidence of fatal cardiac arrest goes from 0.7% to 2.9% (el-Banayosy et al., 1998; Mackay, Powell, Osgathorp, & Rozario, 2002) while the acute myocardial infarction range goes from 2% to 15%. In pediatrics, cardiac arrest is more frequent in patients with complex congenital heart disease perioperative (6%) (Rhodes et al., 1999), and is mainly common during cardiac catheterization (1%) (Odegard et al., 2014) and a frequent cause of death at follow-up (22%) of deaths related to congenital heart disease (Nieminen, Jokinen, & Sairanen, 2007); in addition, pediatric patients myocardial infarction is a rare complication during reimplantation of coronary arteries or surgical trauma event (Heusch, 2013; Heusch et al., 2014).

Ischemic heart disease

The main conditions under which patients may experience an ischemic infarction are ischemic heart disease and cardiopulmonary bypass.

The term "ischemic heart disease" refers to a series of clinical pictures that all share an underlying myocardial ischemia, generated spontaneously. Ischemic heart disease is still a major cause of death and disability worldwide, and is associated with higher health care costs than other chronic diseases. In the United States, 13 million people are affected. More than 6 million suffer from angina pectoris, and more than 7 million have already had a heart attack. Although this disease is increasing in lower classes in both Europe and the United States, mortality was reduced by providing treatment and to better control risk factors regarding prevention. Men constitute 70% of all patients with angina pectoris, and an even greater percentage of them are younger than 50 years.

The main risk factors include genetic predisposition, a high-calorie diet and high in fat, smoking and lack of physical activity, which together with high plasma levels of low-density lipoprotein cholesterol, low levels of high-density lipoproteins, hypertension, diabetes mellitus, age over 75 years, morbid obesity, peripheral myocardial ischemia, and/or prior cerebrovascular disease and myocardial infarction, are all risk factors for coronary atherosclerosis.

The atherosclerotic disease represents the overall main cause of heart disease, being able to determine a more or less marked degree of occlusion of coronary arteries (Goldschmidt-Clermont et al., 2005): its starting point is an alteration of endothelial functions including the regulation of vascular tone, retaining control of adhesion and diapedesis of inflammatory cells. When these functions are lost, it is inadequate to determine vasoconstriction, thrombus formation in the lumen, and abnormal interaction by the vascular endothelium activated with erythrocytes, platelets, and monocytes. These changes in endothelial defensive capabilities lead to the formation of atherosclerotic plaque as a result of the accumulation of fat subintimal smooth muscle cells, fibroblasts, and the intercellular matrix. These collections are distributed unevenly in different portions of the epicardial coronary circulation.

The site of the occlusion influences clinical outcomes and their severity, so when it caused by the main left coronary artery and proximal left anterior descending that can be very dangerous for the patient's life. Very often, in the case of reductions in coronary patency, the creation of collateral circulation may, at rest but not in the case of increased blood flow demand, overcome the myocardial perfusion. Other pathological conditions such as vascular spasms (angina, Prinzmetal's variant), vasomotor physiological phenomena, coronary artery stenosis, coronary arterial thrombi, emboli, abnormal vascular dilating capacity, impaired constrictions (microvascular angina), and far more rare in adult congenital anomalies, may also reduce coronary blood flow and myocardial perfusion.

In addition to inadequate adaptations of coronary blood flow, myocardial ischemia can also occur when oxygen demands are exceedingly increased. Obviously, this happens especially when coronary blood flow is reduced, as in a left ventricular hypertrophy caused by aortic stenosis or hypertensive patient. An alteration in the transport capacity of oxygen by hemoglobin (e.g., under conditions of severe anemia or carbon monoxide poisoning conditions) rarely can determine an ischemic event but can reduce the threshold level needed for moderate coronary obstruction manifests clinically.

Consequences of myocardial ischemia

The consequences of inadequate myocardial perfusion affecting the metabolic, mechanical, and electrical aspects of the heart to alter contractile function of the myocardium, determining areas of hypokinesia or akinesia (visible by echocardiography) more or less extensive or even in severe cases protrusions of the wall, thus leading to a significant impairment of myocardial pump function. Besides the above-mentioned metabolic abnormalities, myocardial ischemia also determines electrical and mechanical alterations: as for the first it basically concerns the phase of ventricular repolarization, whose deterioration is visible to the electrocardiogram as reversal of the T wave and, when more severe, from elevation or ST segment depression, respectively, if it is transmural necrosis or subendocardial infarction. Also, the ischemic myocardial damage leads to cardiac electrical instability that would result in possible isolated premature ventricular beats, tachycardia, or atrial fibrillation.

Most of the relative myocardial ischemia mortality and morbidity is still connected to the remodeling that follows the same; in fact, chronic systolic heart failure and inappropriate remodeling of the left ventricle are crucial determinants of morbidity and long-term clinical outcomes (Konstam, Kramer, Patel, Maron, & Udelson, 2011). Cardiac remodeling is not to be considered solely as a late event in postmyocardial infarction, but is closely related to the starting size of the infarct associated with it. Therefore adequate remodeling following a severe ischemia-reperfusion can greatly improve patient outcomes (Heusch et al., 2014).

The most typical and negative changes of the left ventricle that occur after a myocardial infarction are represented by dilatation, hypertrophy, and morphological change: the hypertrophy is an initial compensatory mechanism that takes place in the remote area of myocardium not subject to heart attack, which is useful if limited in time, since a prolonged myocardial hypertrophy causes a concomitantly increase in the left ventricular mass. The adaptive process of ventricular hypertrophy may be associated with long-term adverse clinical outcomes (Verma et al., 2008). Another feature of left ventricular remodeling is ischemic myocardial interstitial fibrosis. From a molecular aspect at the base of the left ventricular remodeling and its adverse effects, a fundamental role is played by the activation of Akt1/2 and extracellular signal-regulated kinase, that in long term, leads to deleterious hypertrophy of cardiomyocytes. Another of the most important markers of this counterproductive adaptation to the postmyocardial infarction is the activation of a long-term neurohormonal response, in particular the sympathetic nervous system and the Renin angiotensin aldosterone system: both the levels of aldosterone (Guder et al., 2007) and catecholamines (Benedict et al., 1996) are in fact useful predictors of cardiovascular mortality in patients postmyocardial infarction.

Ischemia of large portions of the ventricle may cause a transient ventricular insufficiency, and if 20%−25% of ischemia affects the left ventricular mass, this can lead to heart failure and cardiomegaly, defining a framework of "ischemic cardiomyopathy" that can also occur directly with heart failure without other earlier symptoms.

The involvement of papillary muscles and the remaining apparatus involved in the functioning of the mitral valve can lead to acute mitral regurgitation with possible symptoms of heart failure despite a limited ischemic area.

The course of myocardial ischemia can be stable and progressive, regress to an asymptomatic stage, or determine the sudden death of the patient.

Ischemia-related metabolic alterations

Myocardial ischemia causes important modifications, that usually occur in a temporal sequence defined as the "ischemic cascade," which first involves the metabolic aspect, then the mechanical and finally, the myocardial electric cells.

Unlike in hypoxia, where due to reduced availability of oxygen, can continue the production of glycolytic energy by anaerobic glycolysis, in ischemia, due to the reduced supply of oxygen and nutrients due to a reduced blood flow, it compromises the availability of substrates for glycolysis. Under physiological conditions, the myocardium metabolizes fatty acids (60%−90% of which constitute the total energy source) and glucose (10%−40% of total energy) by transforming them into carbon dioxide and water. During myocardial ischemia that quickly leads to the arrest of aerobic metabolism: fatty acids cannot be oxidized so that, through anaerobic glycolysis, glucose is metabolized to lactic acid resulting in the release of free fatty acids (FFA) causing arrhythmia and inhibiting mitochondrial K-ATP channels (mK-ATP) that can no longer maintain mitochondrial membrane potential, and glycogenolysis is gradually slowed and inhibited by the increase in NADH and FADH2 and the drop in pH (8−10). This is followed by the formation of lactates, a decrease in intracellular pH, the progressive reduction of high-energy phosphates such as ATP and phosphate, as well as the accumulation of several catabolites, including those resulting from the pool of adenine nucleotide. The reduction of the reserves of ATP affects the ion exchange at the level of the sarcolemma, with increased Na^+ and K^+ intracellular reduction due to its outflow, causing the onset of cardiomyocytes death. The increase in intracellular Na^+ results in an increase of intracellular calcium through an increased exchange $Na^+−Ca^{2+}$. The reduced availability of ATP also lowers the intake of sarcoplasmic reticulum Ca^{2+} and reduces the extrusion of Ca^{2+} from the cell. The increase in intracellular Ca^{2+} produces an overload of Ca^{2+} in mitochondria and that further depresses ATP production. The Ca^{2+} is therefore playing a central role in the vicious cycle that leads to irreversible damage of the cell in the case of persistent ischemia. If hypoxia continues, the aggravation of ATP depletion causes more damage: the cytoskeleton is dispersed, resulting in the disappearance of ultrastructural characteristics such as microvilli and the formation of small protrusions on the cell surface. At this stage, the mitochondria are swollen, and due to an inability to control their volume, the endothelial pattern remains dilated, and the whole cell becomes particularly voluminous, with increased water content and the concentration of sodium and potassium chlorides. Due to the high dependence of myocardial function by oxygen, severe ischemia induces loss of contractility within 60 seconds. This can leads to acute heart failure even before myocardial necrosis occurs. If the availability of oxygen is restored, all these changes are reversible. In contrast, if ischemia persists, it will cause irreversible damage and necrosis.

A myocardial ultrastructural alteration, cellular and mitochondrial swelling, and glycogen depletion are potentially reversible; in fact, only one severe ischemia lasting at least 20−30 minutes leads to irreversible damage, with necrosis of cardiomyocytes. The ultrastructural evidence of irreversible myocardial damage occurs only after a serious and prolonged myocardial ischemia. A key characteristic of the initial phases of myocardial necrosis is the loss of the integrity of sarcoplasmic membrane, which allows the intracellular molecules to move to cardiac interstitium and then into circulation.

Necrosis is typically complete within 6 hours after the onset of severe myocardial ischemia.

From a morphological point of view, the irreversible myocardial damage is associated with a serious mitochondrial swelling, extensive cytoplasmic membrane damage with the appearance of myelin figures, and a bulge of lysosomes where in mitochondrial matrix it may develop large amorphous and dense bodies. In the myocardium, these elements attest to the irreversibility of the lesions and can be observed after 30—40 minutes of ischemia.

A fatal injury begins to be determined after 20 minutes of the coronary blood flow, and myocardial exclusion progresses as a wavefront from subendocardial to void the epicardium. After 60 minutes, the third most internal wall is irreversibly damaged. After 3 hours it remains only a fabric edge subepicardial, and the transmural infarct extension is complete after 3—6 hours from occlusion (Kloner & Jennings, 2001).

In contrast, following coronary occlusion in experimental model subendocardial infarction undergoes to irreversible myocardial injury with in an hour and myocardial necrosis progression continued for 4-6 hours. The apoptosis of cardiomyocytes has been identified as a fundamental process of all stages of myocardial infarction, suggesting that it might be largely responsible for the death of cardiomyocytes during the acute phase of myocardial ischemia, as well as the progressive loss of cells survived during the first subacute and chronic stages (Takemura & Fujiwara, 2006); apoptosis also seems to play a significant role in the deterioration of left ventricular function in ischemic regions (Wolf & Green, 1999).

This may suggest that the inhibition of apoptosis may limit the loss of cardiomyocytes induced by ischemia, particularly in chronic ischemic myocardium, reducing left ventricular remodeling and improving the prognosis. Apoptosis is primarily triggered by the liberation of proapoptotic molecules from damaged mitochondria, called the intrinsic or mitochondrial apoptotic pathway. The latter is the result of increased mitochondrial permeability and the release of proapoptotic molecules within the cytoplasm. The release of mitochondrial proteins in the cytoplasm breaks the delicate balance between pro- and antiapoptotic members of the BCL family: under normal conditions, antiapoptotic proteins reside in the cytoplasm and in the mitochondrial membranes where they control the mitochondrial permeability and prevent leakage of mitochondrial proteins. When cells are deprived of survival signals there is activation of damage and stress sensors, also members of the BCL family such as BIM, BID, and BAD. These sensors trigger in turn two effectors proapoptotic proteins BAX and BAK, which go to make up the channels on the mitochondrial membrane by encouraging the passage of proteins in the cytoplasm. The proapoptotic BCL-2 and BCL-X factors are antagonized by the proapoptotic proteins. The combination of all these events leads ultimately to the release of cytochrome into the cytoplasm; a mitochondrial protein can trigger, along with other within a structure called the apoptosome, caspase 9, the initiator of the mitochondrial pathway. A group of cysteine aspartate proteases, which in normal cells reside in the cytosol as inactive form (procaspases). This terminates at the final phase of apoptosis, also common to the intrinsic pathway activation, characterized by activation of effector caspases, which ultimately determine the characteristic DNA cleavage, fragmentation of the nucleus, and lysis of myofibrils. Apoptotic cells and their fragments are then recognized by different receptors on macrophages and finally swallowed. In the ischemic heart then we witness alternative activation and inactivation of multiple genes and their products: among these are found just proapoptotic factors. Among the events resulting from myocardial ischemia are found metabolic acidosis, reduced production of adenosine triphosphate (ATP), the loss of sodium—potassium ATPase pump, as well as the release of chemicals that stimulate chemo- and mechanoreceptors innervated by nonmyelinated nerve cells in the structure of cardiomyocytes and around the coronary arteries (Foreman, Garrett, & Blair, 2015). Among the substances released there are lactate, serotonin, oxygen-free radicals, and adenosine (Benson, Eckert, & McCleskey, 1999; Fu & Longhurst, 2005); in addition to these there are others, such as serotonin, 5-hydroxytryptamine, and thromboxane A2 coming from platelets that are frequently found aggregated in correspondence at the coronary obstruction, which may be partly responsible for myocardial ischemia and angina (Fu, Guo, & Longhurst, 2008; Fu & Longhurst, 2002). In the ischemic heart, there is considerable production of reactive oxygen species (ROS), or oxygen molecules with an excess of electrons that make it chemically very reactive: ROS, using the phenomena of lipid cell membrane constituents, can damage cardiomyocytes and thus contribute to ischemic damage.

Treatment options for myocardial ischemia

Although timely reperfusion of the myocardium at risk represents the most effective way to restore homeostasis of myocardial cell, it is also important to maintain an optimal balance between supply and demand of oxygen, so as to save the widest possible area of myocardium at risk around the areas most damaged by ischemia. The progression of irreversible damage is in fact accelerated by factors that increase myocardial oxygen consumption as in tachycardia or reduce its contribution as in arterial hypotension. Potential mechanisms and measures to counteract the lack of oxygen at myocardial cells include first the reduction of myocardial oxygen consumption: this will lead to prolonged survival by saving energy on the part of the myocardial cells and to a reduction in the catabolite production. Since your heart rate is the main determining factor in myocardial consumption of oxygen (which is halved by halving the number of heartbeats), its reduction is a crucial mode of protection from myocardial ischemia (Lanza, Fox, & Crea, 2006) by keeping the patient at rest, possibly using light sedation, and through the use of beta-blocker medications, which reduce the myocardial work, also reducing blood pressure (after load) and contractility.

Although there are not always unique data to consider about the deleterious effects caused by FFA (Schwartz, Greyson, Wisneski, & Garcia, 1994). It is believed that this metabolic pathway in ischemic conditions presents several disadvantages to the myocardial cells. Indeed, in addition to a higher oxygen consumption, it also involves the inhibition of glucose oxidation and an increase in the production of lactate and protons, which further depresses myocardial function in ischemic areas. Furthermore, the FFA alter cellular ionic homeostasis leading to an increase of arrhythmic substrate. Therefore therapeutic interventions to partially inhibit the uptake and oxidation of FFA and to promote the use of glucose utilization during ischemia should have favorable effects on myocardial cells (Opie, 2008).

A solution of glucose—insulin—potassium (GIK) has long been used with the aim of increasing glycolysis and reducing the uptake of FFA and metabolism of myocardial cells during acute myocardial infraction, and even the administration of insulin alone was proposed with a similar goal. The effects of GIK solution (or insulin alone) in preserving the integrity of cardiomyocytes in this context, however, remain controversial. In fact, some studies (Opie, 2008) have shown favorable effects on survival with this approach, both in diabetic patients and in nondiabetics, but other studies have failed to confirm these findings.

Pre- and postconditioning

The 2—5 minutes typically transient ischemia, it makes metabolically more resistant to any possible subsequent more prolonged ischemia; this phenomenon is called ischemic preconditioning. It provides protection against ischemic episode occurring in 2 hours after transient ischemic preconditioning (early preconditioning), but also against a late-occurring episode after 24 hours and ischemia for up to 72 hours after the episode of ischemic preconditioning (delayed or chronic preconditioning; Vinten-Johansen, Zhao, Jiang, Zatta, & Dobson, 2007). This reduces the size of myocardial necrosis by reducing the infarcted area by 40%—75% and protects the heart from ischemia-induced damage. A key role is carried out in particular by the protein kinase c (PKC), and specifically its isoform PKCε; it seems like a central mediator of the ischemic pre conditioning (IPC). In fact, various substances that accumulate in the interstitium during myocardial ischemia and that are involved in the IPC have in common the ability to activate PKC through the activation of phospholipase c (such as bradykinin, adenosine, norepinephrine, opioid; Cohen & Downey, 2008; Critz, Cohen, & Downey, 2005) or through other routes, such as the activation of protein kinase G (as in the case of nitric oxide; Cohen, Yang, & Downey, 2006) (Fig. 2.1).

Following the IPC, that may result the activation of PKC, seems to be the opening of ATP-sensitive potassium channels (K-ATP channels) and the cytoplasmic membrane of mitochondria is opening, which is mediated

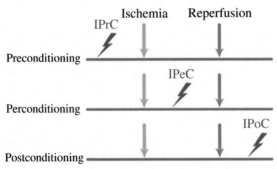

Figure 2.1 Three types of conditioning (Wang et al., 2015).

by phosphorylation of proteins channel by precisely PKC and that causes a reduction of the calcium influx and cellular energy expenditure. The IPC can be blocked by glibenclamide (Cleveland, Meldrum, Cain, Banerjee, & Harken, 1997), an antagonist of K-ATP channels.

The formation of low amounts of ROS after the brief period of ischemia, although insufficient to cause cellular damage, may be able to activate cellular mechanisms involved in the cardioprotection. A low concentration of ROS seems to activate certain enzymatic pathways involved in IPC (guanylate cyclase, cyclooxygenase, tissue factor, and tyrosine; Tritto et al., 1997). Moreover, some substances capable of generating ROS (acetylcholine, bradykinin, opioids, and anesthetics) are also known as preconditioning reagents (McPherson & Yao, 2001).

Finally, recent data suggest that preconditioning effects against ischemic myocardial damage can be achieved by transient episodes of ischemia induced in other areas of the body (e.g., lower limbs; Vinten-Johansen et al., 2005), a phenomenon called remote preconditioning. A final protective mechanism, the postconditioning myocardial infarction, is the ability to provide pharmacological agonists administered by throwing an intermittent ischemia or heart at the time of reperfusion; this protection mechanism by myocardial ischemic injury has the greatest potential to influence the irreversible damage because it can be induced after myocardial ischemia is established, rather than requiring pretreatment (Downey & Cohen, 2006).

References

Benedict, C. R., Shelton, B., Johnstone, D. E., Francis, G., Greenberg, B., Konstam, M., ... Yusuf, S. (1996). Prognostic significance of plasma norepinephrine in patients with asymptomatic left ventricular dysfunction. SOLVD Investigators. *Circulation*, *94*(4), 690−697.

Benson, C. J., Eckert, S. P., & McCleskey, E. W. (1999). Acid-evoked currents in cardiac sensory neurons: A possible mediator of myocardial ischemic sensation. *Circulation Research*, *84*(8), 921−928.

Bradley, S. M., Huszti, E., Warren, S. A., Merchant, R. M., Sayre, M. R., & Nichol, G. (2012). Duration of hospital participation in get with the guidelines-resuscitation and survival of in-hospital cardiac arrest. *Resuscitation*, *83*(11), 1349−1357. Available from https://doi.org/10.1016/j.resuscitation.2012.03.014.

Chan, P. S., Jain, R., Nallmothu, B. K., Berg, R. A., & Sasson, C. (2010). Rapid response teams: A systematic review and meta-analysis. *Archives of Internal Medicine*, *170*(1), 18−26. Available from https://doi.org/10.1001/archinternmed.2009.424.

Chen, L. M., Nallamothu, B. K., Spertus, J. A., Li, Y., Chan, P. S., & American Heart Association's Get With the Guidelines-Resuscitation, I. (2013). Association between a hospital's rate of cardiac arrest incidence and cardiac arrest survival. *JAMA Internal Medicine*, *173*(13), 1186–1195. Available from https://doi.org/10.1001/jamainternmed.2013.1026.

Cleveland, J. C., Jr., Meldrum, D. R., Cain, B. S., Banerjee, A., & Harken, A. H. (1997). Oral sulfonylurea hypoglycemic agents prevent ischemic preconditioning in human myocardium. Two paradoxes revisited. *Circulation*, *96*(1), 29–32.

Cohen, M. V., & Downey, J. M. (2008). Adenosine: Trigger and mediator of cardioprotection. *Basic Research in Cardiology*, *103*(3), 203–215. Available from https://doi.org/10.1007/s00395-007-0687-7.

Cohen, M. V., Yang, X. M., & Downey, J. M. (2006). Nitric oxide is a preconditioning mimetic and cardioprotectant and is the basis of many available infarct-sparing strategies. *Cardiovascular Research*, *70*(2), 231–239. Available from https://doi.org/10.1016/j.cardiores.2005.10.021.

Critz, S. D., Cohen, M. V., & Downey, J. M. (2005). Mechanisms of acetylcholine- and bradykinin-induced preconditioning. *Vascular Pharmacology*, *42*(5-6), 201–209. Available from https://doi.org/10.1016/j.vph.2005.02.007.

Cummins, R. O., Chamberlain, D., Hazinski, M. F., Nadkarni, V., Kloeck, W., Kramer, E., ... Cobbe, S. (1997). Recommended guidelines for reviewing, reporting, and conducting research on in-hospital resuscitation: The in-hospital "Utstein style." American Heart Association. *Annals of Emergency Medicine*, *29*(5), 650–679.

Downey, J. M., & Cohen, M. V. (2006). Reducing infarct size in the setting of acute myocardial infarction. *Progress in Cardiovascular Diseases*, *48*(5), 363–371. Available from https://doi.org/10.1016/j.pcad.2006.02.005.

el-Banayosy, A., Brehm, C., Kizner, L., Hartmann, D., Kortke, H., Korner, M. M., ... Korfer, R. (1998). Cardiopulmonary resuscitation after cardiac surgery: A two-year study. *Journal of Cardiothoracic and Vascular Anesthesia*, *12*(4), 390–392.

Foreman, R. D., Garrett, K. M., & Blair, R. W. (2015). Mechanisms of cardiac pain. *Comprehensive Physiology*, *5*(2), 929–960. Available from https://doi.org/10.1002/cphy.c140032.

Fu, L. W., & Longhurst, J. C. (2002). Activated platelets contribute to stimulation of cardiac afferents during ischaemia in cats: Role of 5-HT(3) receptors. *The Journal of Physiology*, *544*(Pt 3), 897–912.

Fu, L. W., & Longhurst, J. C. (2005). Interactions between histamine and bradykinin in stimulation of ischaemically sensitive cardiac afferents in felines. *The Journal of Physiology*, *565*(Pt 3), 1007–1017. Available from https://doi.org/10.1113/jphysiol.2005.084004.

Fu, L. W., Guo, Z. L., & Longhurst, J. C. (2008). Undiscovered role of endogenous thromboxane A2 in activation of cardiac sympathetic afferents during ischaemia. *The Journal of Physiology*, *586*(13), 3287–3300. Available from https://doi.org/10.1113/jphysiol.2007.148106.

Goldschmidt-Clermont, P. J., Creager, M. A., Losordo, D. W., Lam, G. K., Wassef, M., & Dzau, V. J. (2005). Atherosclerosis 2005: Recent discoveries and novel hypotheses. *Circulation*, *112*(21), 3348–3353. Available from https://doi.org/10.1161/circulationaha.105.577460.

Guder, G., Bauersachs, J., Frantz, S., Weismann, D., Allolio, B., Ertl, G., ... Stork, S. (2007). Complementary and incremental mortality risk prediction by cortisol and aldosterone in chronic heart failure. *Circulation*, *115*(13), 1754–1761. Available from https://doi.org/10.1161/circulationaha.106.653964.

Gwinnutt, C. L., Columb, M., & Harris, R. (2000). Outcome after cardiac arrest in adults in UK hospitals: Effect of the 1997 guidelines. *Resuscitation, 47*(2), 125–135.

Heusch, G. (2013). Cardioprotection: Chances and challenges of its translation to the clinic. *Lancet, 381*(9861), 166–175. Available from https://doi.org/10.1016/s0140-6736(12)60916-7.

Heusch, G., Libby, P., Gersh, B., Yellon, D., Bohm, M., Lopaschuk, G., & Opie, L. (2014). Cardiovascular remodeling in coronary artery disease and heart failure. *Lancet, 383*(9932), 1933–1943. Available from https://doi.org/10.1016/s0140-6736(14)60107-0.

Kloner, R. A., & Jennings, R. B. (2001). Consequences of brief ischemia: Stunning, pre-conditioning, and their clinical implications. *Circulation, 104*(25), 3158.

Konstam, M. A., Kramer, D. G., Patel, A. R., Maron, M. S., & Udelson, J. E. (2011). Left ventricular remodeling in heart failure: Current concepts in clinical significance and assessment. *JACC: Cardiovascular Imaging, 4*(1), 98–108. Available from https://doi.org/10.1016/j.jcmg.2010.10.008.

Lanza, G. A., Fox, K., & Crea, F. (2006). Heart rate: A risk factor for cardiac diseases and outcomes? Pathophysiology of cardiac diseases and the potential role of heart rate slowing. *Advances in Cardiology, 43*, 1–16. Available from https://doi.org/10.1159/000095401.

Mackay, J. H., Powell, S. J., Osgathorp, J., & Rozario, C. J. (2002). Six-year prospective audit of chest reopening after cardiac arrest. *European Journal of Cardio-Thoracic Surgery, 22*(3), 421–425.

McPherson, B. C., & Yao, Z. (2001). Morphine mimics preconditioning via free radical signals and mitochondrial K(ATP) channels in myocytes. *Circulation, 103*(2), 290–295.

Nadkarni, V. M., Larkin, G. L., Peberdy, M. A., Carey, S. M., Kaye, W., Mancini, M. E., ... Berg, R. A. (2006). First documented rhythm and clinical outcome from in-hospital cardiac arrest among children and adults. *JAMA, 295*(1), 50–57. Available from https://doi.org/10.1001/jama.295.1.50.

Nieminen, H. P., Jokinen, E. V., & Sairanen, H. I. (2007). Causes of late deaths after pediatric cardiac surgery: A population-based study. *Journal of the American College of Cardiology, 50*(13), 1263–1271. Available from https://doi.org/10.1016/j.jacc.2007.05.040.

Odegard, K. C., Bergersen, L., Thiagarajan, R., Clark, L., Shukla, A., Wypij, D., & Laussen, P. C. (2014). The frequency of cardiac arrests in patients with congenital heart disease undergoing cardiac catheterization. *Anesthesia & Analgesia, 118*(1), 175–182. Available from https://doi.org/10.1213/ANE.0b013e3182908bcb.

Opie, L. H. (2008). Metabolic management of acute myocardial infarction comes to the fore and extends beyond control of hyperglycemia. *Circulation, 117*(17), 2172–2177. Available from https://doi.org/10.1161/circulationaha.108.780999.

Rhodes, J. F., Blaufox, A. D., Seiden, H. S., Asnes, J. D., Gross, R. P., Rhodes, J. P., ... Rossi, A. F. (1999). Cardiac arrest in infants after congenital heart surgery. *Circulation, 100*(19 Suppl), Ii194–Ii199.

Sandroni, C., Ferro, G., Santangelo, S., Tortora, F., Mistura, L., Cavallaro, F., ... Antonelli, M. (2004). In-hospital cardiac arrest: Survival depends mainly on the effectiveness of the emergency response. *Resuscitation, 62*(3), 291–297. Available from https://doi.org/10.1016/j.resuscitation.2004.03.020.

Schipper, D. A., Marsh, K. M., Ferng, A. S., Duncker, D. J., Laman, J. D., & Khalpey, Z. (2016). The critical role of bioenergetics in donor cardiac allograft preservation. *Journal of Cardiovascular Translational Research, 9*(3), 176−183. Available from https://doi.org/10.1007/s12265-016-9692-2.

Schwartz, G. G., Greyson, C., Wisneski, J. A., & Garcia, J. (1994). Inhibition of fatty acid metabolism alters myocardial high-energy phosphates in vivo. *American Journal of Physiology, 267*(1 Pt 2), H224−H231.

Takemura, G., & Fujiwara, H. (2006). Morphological aspects of apoptosis in heart diseases. *Journal of Cellular and Molecular Medicine, 10*(1), 56−75.

Thygesen, K., Alpert, J. S., Jaffe, A. S., Simoons, M. L., Chaitman, B. R., & White, H. D. (2012). Third universal definition of myocardial infarction. *Global Heart, 7*(4), 275−295. Available from https://doi.org/10.1016/j.gheart.2012.08.001.

Tritto, I., D'Andrea, D., Eramo, N., Scognamiglio, A., De Simone, C., Violante, A., . . . Ambrosio, G. (1997). Oxygen radicals can induce preconditioning in rabbit hearts. *Circulation Research, 80*(5), 743−748.

Verma, A., Meris, A., Skali, H., Ghali, J. K., Arnold, J. M., Bourgoun, M., . . . Solomon, S. D. (2008). Prognostic implications of left ventricular mass and geometry following myocardial infarction: The VALIANT (VALsartan In Acute myocardial iNfarcTion) Echocardiographic Study. *JACC: Cardiovascular Imaging, 1*(5), 582−591. Available from https://doi.org/10.1016/j.jcmg.2008.05.012.

Vinten-Johansen, J., Zhao, Z. Q., Jiang, R., Zatta, A. J., & Dobson, G. P. (2007). Preconditioning and postconditioning: Innate cardioprotection from ischemia-reperfusion injury. *Journal of Applied Physiology (1985), 103*(4), 1441−1448. Available from https://doi.org/10.1152/japplphysiol.00642.2007.

Vinten-Johansen, J., Zhao, Z. Q., Zatta, A. J., Kin, H., Halkos, M. E., & Kerendi, F. (2005). Postconditioning—A new link in nature's armor against myocardial ischemia-reperfusion injury. *Basic Research in Cardiology, 100*(4), 295−310. Available from https://doi.org/10.1007/s00395-005-0523-x.

Wallmuller, C., Meron, G., Kurkciyan, I., Schober, A., Stratil, P., & Sterz, F. (2012). Causes of in-hospital cardiac arrest and influence on outcome. *Resuscitation, 83*(10), 1206−1211. Available from https://doi.org/10.1016/j.resuscitation.2012.05.001.

Wang, Y., Reis, C., Applegate, R., Stier, G., Martin, R., & Zhang, J. H. (2015). Ischemic conditioning-induced endogenous brain protection: Applications pre-, per- or post-stroke. *Experimental Neurology, 272*, 26−40. Available from https://doi.org/10.1016/j.expneurol.2015.04.009.

Wolf, B. B., & Green, D. R. (1999). Suicidal tendencies: Apoptotic cell death by caspase family proteinases. *The Journal of Biological Chemistry, 274*(29), 20049−20052.

Yellon, D. M., & Hausenloy, D. J. (2007). Myocardial reperfusion injury. *New England Journal of Medicine, 357*(11), 1121−1135. Available from https://doi.org/10.1056/NEJMra071667.

CHAPTER 3

Myocardial reperfusion

Abstract

Reperfusion injury, also called ischemia—reperfusion injury or reoxygenation injury, is the tissue damage caused when blood supply returns to tissue (re- + perfusion) after a period of ischemia or lack of oxygen (anoxia or hypoxia). The absence of oxygen and nutrients from blood during the ischemic period creates a condition in which the restoration of circulation results in inflammation and oxidative damage through the induction of oxidative stress rather than (or along with) restoration of normal function. This needs to be addressed timely and effectively for better outcome of cardiac functions.

Keywords: Reperfusion injury; oxidative stress; inflammation

Contents

Myocardial reperfusion

The dependence of myocardial recovery on treatment time applies particularly to patients treated with fibrinolysis or percutaneous coronary intervention (PCI) (De Luca, Suryapranata, Ottervanger, & Antman, 2004). This time dependence can be especially critical for fibrinolysis due to decreased efficacy of fibrinolytic drugs with the progressive organization of coronary thrombi over time. The study conducted by De Luca et al. (2004) demonstrated how every minute of delay at the beginning of the

Pathophysiology of Ischemia Reperfusion Injury and Use of Fingolimod in Cardioprotection
DOI: https://doi.org/10.1016/B978-0-12-818023-5.00003-0

treatment, such as performing a primary angioplasty, impacts on mortality up to 1 year, even after considering the basic conditions of the patient, as can be stated more precisely 7.5% increase in mortality rate for every 30 minutes of delay in initiation of treatment (Fig. 3.1).

The cornerstone of myocardial ischemia treatment consists of procedures enabling rapid restoration of coronary blood flow in the area of the ischemic myocardium (Morel et al., 2016); this is a concept not older than 40 years and was reported by Ross and associates (Ginks et al., 1972; Maroko et al., 1972) who first said that 180 minutes of reperfusion after coronary occlusion reduces the infarct size in dogs. These findings were quickly transferred to patients suffering from acute myocardial infarction who underwent percutaneous coronary intervention (PCI) or thrombolysis to restore blood flow to the myocardium (GISSI, 1986; Ibanez, Heusch, Ovize, & Van de Werf, 2015).

The prevention of cell death by restoring blood flow depends on the severity and duration of preexisting ischemia: the more rapidly blood flow

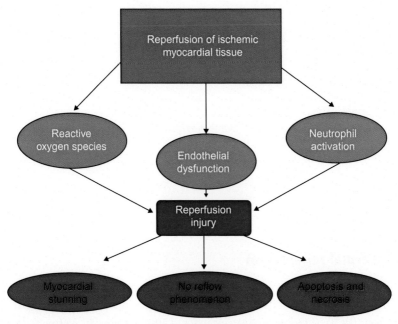

Figure 3.1 Mediators in myocardial reperfusion injury. Reperfusion injury is the outcome of multiple factors that involve the production of reactive oxygen species, endothelial dysfunction, and activation. Reperfusion injury may lead to myocardial stunning, no-reflow phenomenon, and apoptosis and necrosis.

is restored, the greater will be the recovery of left ventricular systolic function, improving diastolic function, and the reduction in overall mortality.

The benefits of coronary reperfusion have already been established Krug, Du Mesnil de, and Korb (1966) and Kloner, Ganote, and Jennings (1974) have shown the drawback in studies conducted on dogs and the so-called "coronary no-reflow phenomenon," which is the absence of reperfusion of the ischemic area with terminal coronary occlusion. According to these studies, 40 minutes ischemia is long enough to cause irreversible damage to most cardiomyocytes, but with little or no change in the pattern of perfusion of large areas of previously ischemic tissue, after 90 minutes of transient ischemia following instead of reduced or absent perfusion of large areas of ischemic tissue. The no-reflow phenomenon (Kishi, Yamada, Okamatsu, & Sunagawa, 2007) *is found in approximately 35% of patients undergoing reperfusion treatments, and its incidence increases with increasing delay in the start of reperfusion and is* associated with a poor prognosis in patients with acute myocardial infarction undergoing reperfusion therapy (Betgem et al., 2015). Also note that the interventional or surgical revascularization may actually, in turn, induce a myocardial infarction collateral to the same procedure.

Despite considerable improvements in cardioprotection intraoperative cardioplegic arrest (cessation transient contractile activity of the heart, produced by infusion in coronary circulation hypothermic conditions solutions based on potassium chloride), which allows heart surgery while protecting the myocardium from ischemic damage due to the interruption of blood flow in the coronary arteries, hypothermia and many more advanced surgical techniques. It is not always possible to use surgical reperfusion through coronary artery bypass grafting (CABG) promptly. Although primary angioplasty compared with thrombolysis is able to get a higher rate of reperfusion in patients "late" with respect to the appropriate time to attack the coronary occlusion, it cannot avoid or prevent the same myocardial necrosis, which is closely related to the duration of the occlusion especially in high-risk patients (Antoniucci et al., 2002; De Luca et al., 2003).

The size of infarcted area turns out to be the main determinant of the long-term mortality of chronic heart failure, and therefore, the ability to limit the extent of the myocardial damage remains one of the main objectives of research (Ibanez et al., 2015).

Physiology of myocardial perfusion

At baseline, myocardial oxygen extraction almost maximum at rest and on average reaches about 75% of the arterial oxygen content (Heusch, 2013), so any increase in the oxygen demands by cardiomyocytes can be satisfied only by a proportional increase in coronary blood flow and oxygen supply, viable firstly by vasodilation of the coronary arterioles (Braunwald, 1971). In addition to the coronary blood flow, oxygen supply is determined directly from the arterial content of oxygen (PaO_2): it is equal to the product of the hemoglobin concentration and arterial oxygen saturation, plus a small amount of oxygen dissolved in plasma.

The basal requirement of myocardial oxygen is low (15%), and the cost of activation power is insignificant when the mechanical contraction ceases during diastolic shutdown (as in cardioplegia) and decreases during ischemia. The myocardial oxygen demand is assessed clinically by multiplying heart rate-systolic pressure, resulting in the so-called "double" as myocardial contractility and wall stress cannot be measured in clinical practice. Normally the myocardium regulates the flow of oxygenated blood depending on the required variables in oxygen, to prevent any hypoperfusion of cardiomyocytes and subsequent ischemia to necrosis: a metabolic adjustment takes place in the course of physical activity or emotional distress, where the increased oxygen demands and energy substrates affect coronary vascular resistance; an autoregulation occurs instead of the physiological changes in blood pressure to maintain within a normal range the coronary flow.

The main determinants of myocardial oxygen consumption (myocardial oxygen demand, MVO_2) are essentially the heart rate, left ventricular contractility and the systolic pressure (or wall of the stressed myocardium). To meet the myocardial oxygen demands is in turn determined by appropriate respiratory function, levels of oxygen saturation, hemoglobin concentration by appropriate coronary circulation. The blood flow through the coronary arteries to the myocardium is influenced by the diameter and vascular tone, by the presence of collateral circulation, from the perfusion pressure determined by the pressure gradient between the aorta and coronary arteries and between arteries and capillaries endocardiac (being the direct flow from endocardium to epicardium), heart rate, which influences the duration of diastole in inverse proportion. Therefore the heart rate is a determinant of both demand and supply of oxygen.

Extra corporeal circulation

Extra corporeal circulation (ECC) is essential for cardiopulmonary bypass in the course of a cardiac surgery and extracorporeal membrane oxygenation or membranes in the course of the extracorporeal cardiopulmonary support (Borgermann, Scheubel, Simm, Silber, & Friedrich, 2007). In the course of ECC, a systemic inflammatory response develops, which contributes to lung and kidney damage. Negative outcomes borne by these and other organs are supported by different mechanisms, such as surgical trauma, the hemodilution, endothelial damage induced by edema, and ischemia—reperfusion injury (IRI) borne by these organs and blood components.

In the course of surgery with ECC, myocardial ischemic damage occurs and can be divided into three main phases: an antecedent ischemia, as a result of a preexisting coronary artery disease; hypotension or ventricular fibrillation ("unprotected ischemia"), and "protected ischemia" which begins with the infusion of cardioplegic solution and connection of cardiopulmonary bypass; then reperfusion injury during intermittent infusions of solution, after removal of the aortic clamp, or after discontinuation of cardiopulmonary bypass. Reperfusion injury may itself be divided into early (<4 hours) and late stage (from 4 to 6 hours).

Role of coronary circulation

The blood flow in the coronary arteries is most represented during the filling phase and diastolic relaxation. The overall resistance of coronary vascular bed can vary considerably, with a concomitant change in the blood flow and myocardial demand, while drawing a constant and high (70%) amount of oxygen from the blood. The coronary vascular resistance is primarily determined by the arterioles and intramyocardial capillaries, while large epicardial arteries are less influenced.

The coronary vessels are the arteries' so-called "functional end," meaning that there are anastomotic branches but they are insufficient to establish a side circle functionally valid. Therefore the area of coronary artery perfusion of a given place distal to the occlusion site represents the region at greater risk of myocardial infarction (MI). When coronary blood flow is reduced significantly for a time ranging between 20 and 40 minutes, myocardial infarction begins to develop from the center of the area at risk, from the subendocardial layer, and then to subepicardial layers and at the edge of the area at risk with the passage of time.

The evolution of myocardial infarction reflects the distribution of coronary blood flow pattern, stream most represented in the outer layers of the myocardium and the margins of the area at risk (Reimer & Jennings, 1979; Reimer, Lowe, Rasmussen, & Jennings, 1977). The evolution of myocardial infarction also depends on the species concerned and the presence or absence of efficient collateral circulation; for example, rodents have a high heart rate and tend to have a rapid evolution of myocardial necrosis.

Among the various aspects that can influence the development of myocardial infarction, in contrast to previous knowledge, the hemodynamic conditions of the subject seem to play a vital role; only the heart rate, to some extent, can influence the progression of myocardial necrosis (Heusch, 2008).

The primates possess few collateral circulation innately but are relatively resistant to ischemic myocardial infarction. In the absence of myocardial infarction coronary occlusion between 40 and 60 minutes, and even after the 90 minutes the size of the infarcted area still smaller in species like in pigs. In humans even after a period of coronary occlusion between 4 and 6 hours, from 20% to 50% of the area at risk remains vital and thus can be retrieved through the reperfusion. The myocardium can be "saved" even after 12 hours of clinical onset, resulting in improvement of the prognosis of the patient (Ndrepepa, Kastrati, Mehilli, Antoniucci, & Schomig, 2009; Schomig et al., 2005). It is unclear presently that what is the mechanism in humans for this ability to resist against blood flow deprivation for so long, either if the presence of well-developed collateral circulation at the time of the stroke, or an inherent resistance to ischemic damage, or if it reflects a preconditioning ischemia—reperfusion due to previous episodes of ischemia reperfusion. Also, the use of medications such as beta blockers, inhibitors of renin-angiotensin system, statins or P2Y12 receptor antagonists, can provide preexisting cardioprotection and thus reduce negative outcomes of myocardial IRI.

Myocardial stunning and hibernation

The extent and nature of myocardial damage depend largely on how long and how much coronary blood flow has been reduced (Ibanez et al., 2015; Skyschally, Schulz, & Heusch, 2008). In fact, a cessation of coronary blood flow lasting less than 20 minutes results in reversible damage represented by the contractile dysfunction associated with a complete although slow recovery after myocardial reperfusion, a phenomenon referred to as "myocardial stunning" (Braunwald & Kloner, 1982; Heyndrickx, Millard, McRitchie, Maroko, & Vatner, 1975). The mechanism underlying this prolonged contractile dysfunction comprises firstly an increased production of reactive oxygen species (ROS), which occurs early during reperfusion (Bolli & Marban, 1999), and an alteration of coupling between excitation and contraction of cardiomyocytes due to oxidative modification of sarcoplasmic reticulum and contractile proteins (Fig. 3.2).

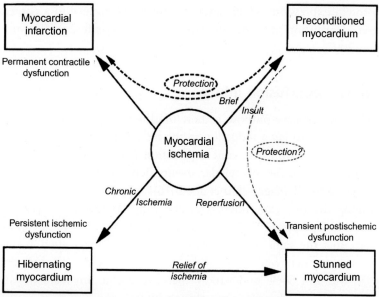

Figure 3.2 Myocardial ischemia may present as reversible myocardial dysfunction in association with reversible electrocardiographic changes and angina pectoris. Additional consequences of myocardial ischemia include: (1) myocardial infarction with permanent contractile dysfunction caused by severe long lasting ischemia, (2) hibernating myocardium as a result of chronic hypoperfusion, (3) stunned myocardium after reperfusion of ischemic myocardial tissue, and (4) preconditioned myocardium after brief ischemic insults. The preconditioned myocardium may in turn offer protection during a subsequent ischemic bout and has been demonstrated to reduce infarct size in animal models. It is not clear at present weather the preconditioned myocardium is capable of having the recovery of stunned myocardium. *Adapted from Vroom, M. B., & van Wezel, H. B. (1996). Myocardial stunning, hibernation, and ischemic preconditioning.* Journal of Cardiothoracic and Vascular Anesthesia, 10 (6), 789–799. doi: http://dx.doi.org/10.1016/S1053-0770(96)80209-6 (Vroom & van Wezel, 1996).

Another phenomenon that follows repeated brief episodes of coronary artery occlusion or the reduction of moderate coronary blood flow but more prolonged is the so-called "hybernating myocardium," which consists of a reduced contractile function of the myocardium that remains vital and subject to a possible recovery following the reperfusion. Although myocardial ischemia was once thought to result in irreversible cellular damage, it is now demonstrated that in cardiac tissue, submitted to the stress of oxygen and substrate deprivation, endogenous mechanisms of cell survival may be activated. These molecular mechanisms result in physiological conditions of adaptation to ischemia, known as myocardial stunning and hibernation. These conditions result from a switch in gene and protein expression, which sustains cardiac cell survival in a context of oxygen deprivation and during the stress of reperfusion (Depre and Vatner, 2007).

Ischemia—reperfusion injury

Although the primary purpose is therefore the timely restoration of coronary blood flow to get a reduction in the infarct size and an improvement in ventricular function, the same reperfusion may result in tissue damage and lethal and a series of cellular events defined as "myocardial ischemia—reperfusion damage" (Hu et al., 2016).

Among the first experimental evidence by Reimer et al. (1977) and Reimer and Jennings (1979) demonstrated in numerous studies conducted on dogs, that the signs of irreversible myocardial damage such as rupture of the sarcolemma, were particularly evident during reperfusion; at that time it was not well understood that irreversible myocardial damage mainly caused by ischemia or reperfusion injury.

The research done by Przyklenk (1997) opened so much debate about the identity of this lethal reperfusion myocardial damage. Finally, an epilogue found by Vinten Johansen is co-author of Zhao et al. (2003), in which they identified the so-called "ischemic postconditioning phenomenon": repeated episodes of coronary occlusion realized early during reperfusion proved to be able to reduce the size of infarcted area in dogs; this discovery was later confirmed in patients with acute myocardial infarction undergoing coronary reperfusion (Staat et al., 2005). These studies once again emphasized what had already been partially mentioned in previous studies on reperfusion, or that modified reperfusion procedures were able to attenuate the myocardial damage may be irreversible (Okamoto, Allen,

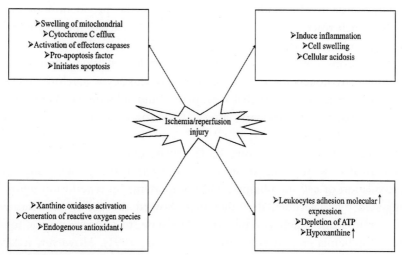

Figure 3.3 Pathophysiology of reperfusion injury.

Buckberg, Bugyi, & Leaf, 1986), reducing the signs of damage at both morphological and functional level (Fig. 3.3) (Musiolik et al., 2010).

Oxygen free radicals

A key role in IRI is played by the release of ROS (Brookes, Yoon, Robotham, Anders, & Sheu, 2004; Halestrap & Richardson, 2015).

During prolonged ischemia, increased levels of purines derived from the catabolism of adenosine triphosphate (ATP), as well as the conversion of Xanthine dehydrogenase in xanthine oxidase by Ca^{2+}, enzyme that converts xanthine to uric acid, during reperfusion the oxygen supply enables xanthine to oxidase to form uric acid from purine accumulated; as a byproduct of the reaction form ROS.

ROS such as the superoxide anion (O_2^-), the hydrogen peroxide (H_2O_2), and the hydroxyl radical (OH) play a key role in both mitochondrial and cytosolic levels as well as for normal functions of mitochondria.

An imbalance between production and removal of free oxygen radicals results in so-called "oxidative stress," involved in various pathological conditions as well as in determining macromolecular damage (Ray, Huang, & Tsuji, 2012). During I/R, evolution of damage by inducing endothelial cell surface alterations by adhesion and activation of circulating neutrophils. The neutrophils release ROS and hydrolytic enzymes that damage cells in ischemia. The ischemia reperfusion also activates the enzyme nitric oxide synthase, which leads to the production of nitric oxide, which reacts with the ROS forming toxic reactive species (peroxynitrite). ROS generated during reperfusion phase cause by various oxidative reactions, including the Nicotinamide adenine dinucleotide phosphate hydrogen (NADPH) oxidase and leukocyte myeloperoxidase, contributing in the endothelial dysfunction and subsequent altered vasomotor capacity of coronary artery (Ambrosio et al., 1993; Gross, O'Rourke, Pelc, & Warltier, 1992).

In addition to these actions, ROS can produce deleterious effects on myocardial cells, resulting in structural and enzymatic proteins and lipid peroxidation of cell membranes. Their importance in reperfusion injury is suggested by experimental studies that showed the ability of treatment with antioxidants, such as superoxide dismutase, to improve cardiac function in ischemia−reperfusion (Ambrosio et al., 1991). However, it lacks the evidence that antioxidant treatment improves myocardial damage associated with reperfusion in humans and studies have not been able to confirm the induction of irreversible damage by free radicals in patients undergoing cardioplegic arrest and reperfusion during coronary artery bypass graft surgery (Milei et al., 2007).

In some pathological conditions, the overproduction of ROS can be stimulated by an overload of calcium (Ca^{2+}), in spite of calcium itself having numerous positive functions at the mitochondrial level (Brookes et al., 2004).

This malicious calcium overload can be determined by the loss of ATP loss in turn determined by an altered function of the Na^+ antiporter/Ca^{2+}.

Inflammation and activation of complement system

The inflammatory damage in the reperfusion is primarily determined by a massive circulating neutrophil invoked in the reperfused area by cytokines and adhesive molecules expressed by endothelial and parenchymal cells but, above all, by the release of cytosolic components from necrotic cells. This inflammation causes, in turn, further tissue damage. The damage triggered by reperfusion is primarily associated with the release of several inflammatory molecules such as cytokines such as IL-1ra, IL-6, IL-8, IL-10, complement activation, and the release of polymorphonuclear leukocytes as well as its actual cellular damage (Bottiger et al., 1996; Madathil et al., 2016).

In the blood of non-survivors following a cardiac arrest a level of IL-6 were significantly higher than patients who survived (Adrie et al., 2002). This state of inflammation triggers very quickly as soon as coronary blood flow is reestablished (Adrie et al., 2004; Niemann et al., 2009). The extent of damage at the cellular level depends on factors, such as the duration of absence of blood flow level particularly in myocardial tissues, the degree of response to the more or less optimal treatment as well as the probability of individual patient survival (Vaahersalo et al., 2014).

To emphasize how many patients with early hemodynamic dysfunction following a cardiopulmonary resuscitation for cardiac arrest have favorable neurological outcome (Chang et al., 2007), therefore the patients suffering from the poor clinical outcomes associated with IRI could be significantly reduced by all those treatments to reduce premature mortality and damage associated with inflammation and trauma present in the chain of ischemia–reperfusion events.

The IL-6, TNF-α, and endothelin cause vasoconstriction, increasing the adhesion of neutrophils and platelets to endothelium, chemotaxis of neutrophils, leading to systemic abnormalities of vascular function. Although the complement system is capable of contributing to IRI (Riedemann & Ward, 2003), some IgM antibodies for unknown reasons tend to be deposited in ischemic tissues; when blood flow is restored, the complement fractions bind to these antibodies, are activated, and induce further cellular damage and increased inflammatory reaction (Zhang et al., 2006).

Increased permeability of mitochondrial permeability transiting pore

Under hypoxic conditions, cellular metabolism switches from a state to an oxidative glycolytic state, the pH is reduced leading to a state of acidosis, and antiporter Na^+/H^+ (6 isoform NHE) tries to reestablish a physiological cellular pH. The sodium gradient thus increases driving the antiporter, Na^+/Ca^{2+} in its function, but can not replace the activity of ATP to prevent its activation. Consequently, it is to determine an excess which eventually leads both to an overproduction of ROS and the opening of the mitochondrial permeability transiting pore (mPTP). Specifically, overstressing of calcium induced by calcium, combined with certain pathological conditions, results in a persistent state of the opening of these mitochondrial pores: among these tissue ischemia, the pathological condition characterized by a reduced value for adenosine triphosphate/adenosine monophosphate (ATP/AMP) and the depletion of adenylic nucleotides.

Once you open the mPTP, mitochondrial membrane potential is no longer maintained, leading to a major influx of water, mitochondrial swelling, and rupture (Brookes et al., 2004).

Specifically, the opening of mitochondrial transition pore determines cell membrane depolarization and a bulge in the array; this, in turn, leads to the outer mitochondrial membrane rupture and release of proteins such as cytochrome C from intermembrane space to the cytosol. The latter cellular event plays an important role in cell death both at the neuronal and myocardial levels (Blomgren, Zhu, Hallin, & Hagberg, 2003; Miura & Tanno, 2012).

The continuing opening of mPTP not only makes the mitochondrion unable to produce ATP through oxidative phosphorylation but also determines the breakdown of ATP molecules produced through glycolysis in an attempt to restore the physiological cellular pH and concentration gradient normally present across the mitochondrial membrane. During prolonged period of ischemia that determine acidosis exerts an inhibition effect on the mitochondrial membrane transition pore opening. Once blood flow is restored and adequate partial oxygen tension reestablished, the cell reverts to aerobic metabolism and the physiological pH increases, resulting in the opening of mPTP and an increased release of ROS. Although as just seen low pH levels can inhibit the opening of mPTP, normalizing pH alone cannot prevent the tissue damage that ensues. Nevertheless, it is still important for the purpose of inhibiting the opening of these mitochondrial pores and ultimately reduce ischemia reperfusion to maintain proper mitochondrial function and prevent pathological conditions predisposing tissue damage related to mitochondria.

Role of mitochondrial DNA

During the development of IRI following the rupture of mitochondria, mitochondrial DNA is released; just considering this process, mitochondrial DNA was recently evaluated as a possible marker of myocardial infarction (Wang et al., 2015; Yue et al., 2015). In fact, in addition to apoptosis induced by the rupture of mitochondria, the same circulating mitochondrial DNA seems to be able to contribute to the death of cardiomyocytes (Yue et al., 2015).

When there is myocardial tissue damage, mitochondrial DNA that is released into the circulation as a result of disruption of mitochondria is recognized by Toll-like receptor 9 (TLR-9) as foreign material, very similar to bacterial DNA; that may develops a systemic inflammatory response (Zhang et al., 2010). In support of this, recent clinical studies have used increased circulating levels of mitochondrial DNA and overexpression of TLR9 as predictors of mortality in patients hospitalized in intensive care unit (Krychtiuk et al., 2015). Furthermore, urinary mitochondrial DNA has been used as a marker of mitochondrial dysfunction in acute kidney injury (Whitaker et al., 2015). Thus circulating mitochondrial DNA could constitute an important means for evaluating myocardial damage, specifically in the transplant ischemia induced by the sequence of events and reperfusion.

In a recent study of coronary artery bypass surgery (CABG), an intervention that temporarily blocks blood flow to the myocardium was seen as being induced the release of massive amounts of mitochondrial DNA free in the circle (Qin et al., 2015), strengthening the hypothesis that this same DNA can be used as markers of ischemia reperfusion.

Myocardial cellular death

Ischemia reperfusion may cause different forms of cell death, such as programmed cell death, necrosis, and apoptosis.

In mammalian cells, there are two most important pathways in the activation of the apoptotic cascade; intrinsic and extrinsic, although there are other lesser known pathways.

The "intrinsic" is triggered by ischemia/reperfusion, hypoxia, and oxidative stress, and is mediated by the damaged mitochondria, which releases substances that activate the caspases cascade and translocation in the nucleus where they induce, directly and indirectly, DNA fragmentation. The "extrinsic" apoptotic pathway entails the death-receptor pathway, triggered by members of the death-receptor superfamily, such as the Fas receptor or the tumor necrosis factor α receptor (TNFR).

References

Adrie, C., Adib-Conquy, M., Laurent, I., Monchi, M., Vinsonneau, C., Fitting, C., ... Cavaillon, J. M. (2002). Successful cardiopulmonary resuscitation after cardiac arrest as a "sepsis-like" syndrome. *Circulation*, *106*(5), 562–568.

Adrie, C., Laurent, I., Monchi, M., Cariou, A., Dhainaou, J. F., & Spaulding, C. (2004). Postresuscitation disease after cardiac arrest: A sepsis-like syndrome? *Current Opinion in Critical Care, 10*(3), 208−212.

Ambrosio, G., Flaherty, J. T., Duilio, C., Tritto, I., Santoro, G., Elia, P. P., . . . Chiariello, M. (1991). Oxygen radicals generated at reflow induce peroxidation of membrane lipids in reperfused hearts. *Journal of Clinical Investigation, 87*(6), 2056−2066. Available from https://doi.org/10.1172/jci115236.

Ambrosio, G., Zweier, J. L., Duilio, C., Kuppusamy, P., Santoro, G., Elia, P. P., et al. (1993). Evidence that mitochondrial respiration is a source of potentially toxic oxygen free radicals in intact rabbit hearts subjected to ischemia and reflow. *The Journal of Biological Chemistry, 268*(25), 18532−18541.

Antoniucci, D., Valenti, R., Migliorini, A., Moschi, G., Trapani, M., Buonamici, P., . . . Santoro, G. M. (2002). Relation of time to treatment and mortality in patients with acute myocardial infarction undergoing primary coronary angioplasty. *American Journal of Cardiology, 89*(11), 1248−1252.

Betgem, R. P., de Waard, G. A., Nijveldt, R., Beek, A. M., Escaned, J., & van Royen, N. (2015). Intramyocardial haemorrhage after acute myocardial infarction. *Nature Reviews Cardiology, 12*(3), 156−167. Available from https://doi.org/10.1038/nrcardio.2014.188.

Blomgren, K., Zhu, C., Hallin, U., & Hagberg, H. (2003). Mitochondria and ischemic reperfusion damage in the adult and in the developing brain. *Biochemical and Biophysical Research Communications, 304*(3), 551−559.

Bolli, R., & Marban, E. (1999). Molecular and cellular mechanisms of myocardial stunning. *Physiological Reviews, 79*(2), 609−634.

Borgermann, J., Scheubel, R. J., Simm, A., Silber, R. E., & Friedrich, I. (2007). Inflammatory response in on- versus off-pump myocardial revascularization: Is ECC really the culprit? *The Thoracic and Cardiovascular Surgery, 55*(8), 473−480. Available from https://doi.org/10.1055/s-2007-965631.

Bottiger, B. W., Bohrer, H., Boker, T., Motsch, J., Aulmann, M., & Martin, E. (1996). Platelet factor 4 release in patients undergoing cardiopulmonary resuscitation—Can reperfusion be impaired by platelet activation? *Acta Anaesthesiologica Scandinavica, 40*(5), 631−635.

Braunwald, E. (1971). Control of myocardial oxygen consumption: Physiologic and clinical considerations. *American Journal of Cardiology, 27*(4), 416−432.

Braunwald, E., & Kloner, R. A. (1982). The stunned myocardium: Prolonged, postischemic ventricular dysfunction. *Circulation, 66*(6), 1146−1149.

Brookes, P. S., Yoon, Y., Robotham, J. L., Anders, M. W., & Sheu, S. S. (2004). Calcium, ATP, and ROS: A mitochondrial love-hate triangle. *American Journal of Physiology − Cell Physiology, 287*(4), C817−C833. Available from https://doi.org/10.1152/ajpcell.00139.2004.

Chang, W. T., Ma, M. H., Chien, K. L., Huang, C. H., Tsai, M. S., Shih, F. Y., . . . Chen, W. J. (2007). Postresuscitation myocardial dysfunction: Correlated factors and prognostic implications. *Intensive Care Medicine, 33*(1), 88−95. Available from https://doi.org/10.1007/s00134-006-0442-9.

De Luca, G., Suryapranata, H., Ottervanger, J. P., & Antman, E. M. (2004). Time delay to treatment and mortality in primary angioplasty for acute myocardial infarction: Every minute of delay counts. *Circulation, 109*(10), 1223−1225. Available from https://doi.org/10.1161/01.cir.0000121424.76486.20.

De Luca, G., Suryapranata, H., Zijlstra, F., van 't Hof, A. W., Hoorntje, J. C., Gosselink, A. T., ... de Boer, M. J. (2003). Symptom-onset-to-balloon time and mortality in patients with acute myocardial infarction treated by primary angioplasty. *Journal of the American College of Cardiology, 42*(6), 991−997.

Depre, C., & Vatner, S. F. (2007). Cardioprotection in stunned and hibernating myocardium. *Heart Failure Reviews, 12,* 307−317.

Ginks, W. R., Sybers, H. D., Maroko, P. R., Covell, J. W., Sobel, B. E., & Ross, J., Jr. (1972). Coronary artery reperfusion. II. Reduction of myocardial infarct size at 1 week after the coronary occlusion. *Journal of Clinical Investigation, 51*(10), 2717−2723. Available from https://doi.org/10.1172/jci107091.

GISSI. (1986). Effectiveness of intravenous thrombolytic treatment in acute myocardial infarction. Gruppo Italiano per lo Studio della Streptochinasi nell'Infarto Miocardico (GISSI). *The Lancet, 1*(8478), 397−402.

Gross, G. J., O'Rourke, S. T., Pelc, L. R., & Warltier, D. C. (1992). Myocardial and endothelial dysfunction after multiple, brief coronary occlusions: Role of oxygen radicals. *American Journal of Physiology, 263*(6 Pt 2), H1703−H1709.

Halestrap, A. P., & Richardson, A. P. (2015). The mitochondrial permeability transition: A current perspective on its identity and role in ischaemia/reperfusion injury. *Journal of Molecular and Cellular Cardiology, 78,* 129−141. Available from https://doi.org/10.1016/j.yjmcc.2014.08.018.

Heusch, G. (2008). Heart rate in the pathophysiology of coronary blood flow and myocardial ischaemia: Benefit from selective bradycardic agents. *British Journal of Pharmacology, 153*(8), 1589−1601. Available from https://doi.org/10.1038/sj.bjp.0707673.

Heusch, G. (2013). Cardioprotection: Chances and challenges of its translation to the clinic. *The Lancet, 381*(9861), 166−175. Available from https://doi.org/10.1016/s0140-6736(12)60916-7.

Heyndrickx, G. R., Millard, R. W., McRitchie, R. J., Maroko, P. R., & Vatner, S. F. (1975). Regional myocardial functional and electrophysiological alterations after brief coronary artery occlusion in conscious dogs. *Journal of Clinical Investigation, 56*(4), 978−985. Available from https://doi.org/10.1172/jci108178.

Hu, T., Wei, G., Xi, M., Yan, J., Wu, X., Wang, Y., ... Wen, A. (2016). Synergistic cardioprotective effects of Danshensu and hydroxysafflor yellow A against myocardial ischemia-reperfusion injury are mediated through the Akt/Nrf2/HO-1 pathway. *International Journal of Molecular Medicine, 38*(1), 83−94. Available from https://doi.org/10.3892/ijmm.2016.2584.

Ibanez, B., Heusch, G., Ovize, M., & Van de Werf, F. (2015). Evolving therapies for myocardial ischemia/reperfusion injury. *Journal of the American College of Cardiology, 65* (14), 1454−1471. Available from https://doi.org/10.1016/j.jacc.2015.02.032.

Kishi, T., Yamada, A., Okamatsu, S., & Sunagawa, K. (2007). Percutaneous coronary arterial thrombectomy for acute myocardial infarction reduces no-reflow phenomenon and protects against left ventricular remodeling related to the proximal left anterior descending and right coronary artery. *International Heart Journal, 48*(3), 287−302.

Kloner, R. A., Ganote, C. E., & Jennings, R. B. (1974). The "no-reflow" phenomenon after temporary coronary occlusion in the dog. *Journal of Clinical Investigation, 54*(6), 1496−1508. Available from https://doi.org/10.1172/jci107898.

Krug, A., Du Mesnil de, R., & Korb, G. (1966). Blood supply of the myocardium after temporary coronary occlusion. *Circulation Research, 19*(1), 57−62.

Krychtiuk, K. A., Ruhittel, S., Hohensinner, P. J., Koller, L., Kaun, C., Lenz, M., ... Speidl, W. S. (2015). Mitochondrial DNA and toll-like receptor-9 are associated with mortality in critically ill patients. *Critical Care Medicine*, *43*(12), 2633–2641. Available from https://doi.org/10.1097/ccm.0000000000001311.

Madathil, R. J., Hira, R. S., Stoeckl, M., Sterz, F., Elrod, J. B., & Nichol, G. (2016). Ischemia reperfusion injury as a modifiable therapeutic target for cardioprotection or neuroprotection in patients undergoing cardiopulmonary resuscitation. *Resuscitation*, *105*, 85–91. Available from https://doi.org/10.1016/j.resuscitation.2016.04.009.

Maroko, P. R., Libby, P., Ginks, W. R., Bloor, C. M., Shell, W. E., Sobel, B. E., & Ross, J., Jr. (1972). Coronary artery reperfusion. I. Early effects on local myocardial function and the extent of myocardial necrosis. *Journal of Clinical Investigation*, *51*(10), 2710–2716. Available from https://doi.org/10.1172/jci107090.

Milei, J., Forcada, P., Fraga, C. G., Grana, D. R., Iannelli, G., Chiariello, M., ... Ambrosio, G. (2007). Relationship between oxidative stress, lipid peroxidation, and ultrastructural damage in patients with coronary artery disease undergoing cardioplegic arrest/reperfusion. *Cardiovascular Research*, *73*(4), 710–719. Available from https://doi.org/10.1016/j.cardiores.2006.12.007.

Miura, T., & Tanno, M. (2012). The mPTP and its regulatory proteins: Final common targets of signalling pathways for protection against necrosis. *Cardiovascular Research*, *94*(2), 181–189. Available from https://doi.org/10.1093/cvr/cvr302.

Morel, S., Christoffersen, C., Axelsen, L. N., Montecucco, F., Rochemont, V., Frias, M. A., ... Kwak, B. R. (2016). Sphingosine-1-phosphate reduces ischaemia-reperfusion injury by phosphorylating the gap junction protein Connexin43. *Cardiovascular Research*, *109*(3), 385–396. Available from https://doi.org/10.1093/cvr/cvw004.

Musiolik, J., van Caster, P., Skyschally, A., Boengler, K., Gres, P., Schulz, R., & Heusch, G. (2010). Reduction of infarct size by gentle reperfusion without activation of reperfusion injury salvage kinases in pigs. *Cardiovascular Research*, *85*(1), 110–117. Available from https://doi.org/10.1093/cvr/cvp271.

Ndrepepa, G., Kastrati, A., Mehilli, J., Antoniucci, D., & Schomig, A. (2009). Mechanical reperfusion and long-term mortality in patients with acute myocardial infarction presenting 12 to 48 hours from onset of symptoms. *JAMA*, *301*(5), 487–488. Available from https://doi.org/10.1001/jama.2009.32.

Niemann, J. T., Rosborough, J. P., Youngquist, S., Shah, A. P., Lewis, R. J., Phan, Q. T., & Filler, S. G. (2009). Cardiac function and the proinflammatory cytokine response after recovery from cardiac arrest in swine. *Journal of Interferon & Cytokine Research*, *29*(11), 749–758. Available from https://doi.org/10.1089/jir.2009.0035.

Okamoto, F., Allen, B. S., Buckberg, G. D., Bugyi, H., & Leaf, J. (1986). Reperfusion conditions: Importance of ensuring gentle versus sudden reperfusion during relief of coronary occlusion. *Journal of Thoracic and Cardiovascular Surgery*, *92*(3 Pt 2), 613–620.

Przyklenk, K. (1997). Lethal myocardial "reperfusion injury": The opinions of good men. *Journal of Thrombosis and Thrombolysis*, *4*(1), 5–6.

Qin, C., Liu, R., Gu, J., Li, Y., Qian, H., Shi, Y., & Meng, W. (2015). Variation of perioperative plasma mitochondrial DNA correlate with peak inflammatory cytokines caused by cardiac surgery with cardiopulmonary bypass. *Journal of Cardiothoracic Surgery*, *10*, 85. Available from https://doi.org/10.1186/s13019-015-0298-6.

Ray, P. D., Huang, B. W., & Tsuji, Y. (2012). Reactive oxygen species (ROS) homeostasis and redox regulation in cellular signaling. *Cell Signaling, 24*(5), 981–990. Available from https://doi.org/10.1016/j.cellsig.2012.01.008.

Reimer, K. A., & Jennings, R. B. (1979). The "wavefront phenomenon" of myocardial ischemic cell death. II. Transmural progression of necrosis within the framework of ischemic bed size (myocardium at risk) and collateral flow. *Laboratory Investigation, 40* (6), 633–644.

Reimer, K. A., Lowe, J. E., Rasmussen, M. M., & Jennings, R. B. (1977). The wavefront phenomenon of ischemic cell death. 1. Myocardial infarct size vs duration of coronary occlusion in dogs. *Circulation, 56*(5), 786–794.

Riedemann, N. C., & Ward, P. A. (2003). Complement in ischemia reperfusion injury. *The American Journal of Pathology, 162*(2), 363–367. Available from https://doi.org/10.1016/s0002-9440(10)63830-8.

Schomig, A., Mehilli, J., Antoniucci, D., Ndrepepa, G., Markwardt, C., Di Pede, F., ... Kastrati, A. (2005). Mechanical reperfusion in patients with acute myocardial infarction presenting more than 12 hours from symptom onset: A randomized controlled trial. *JAMA, 293*(23), 2865–2872. Available from https://doi.org/10.1001/jama.293.23.2865.

Skyschally, A., Schulz, R., & Heusch, G. (2008). Pathophysiology of myocardial infarction: Protection by ischemic pre- and postconditioning. *Herz, 33*(2), 88–100. Available from https://doi.org/10.1007/s00059-008-3101-9.

Staat, P., Rioufol, G., Piot, C., Cottin, Y., Cung, T. T., L'Huillier, I., ... Ovize, M. (2005). Postconditioning the human heart. *Circulation, 112*(14), 2143–2148. Available from https://doi.org/10.1161/circulationaha.105.558122.

Vaahersalo, J., Skrifvars, M. B., Pulkki, K., Stridsberg, M., Rosjo, H., Hovilehto, S., ... Ruokonen, E. (2014). Admission interleukin-6 is associated with post resuscitation organ dysfunction and predicts long-term neurological outcome after out-of-hospital ventricular fibrillation. *Resuscitation, 85*(11), 1573–1579. Available from https://doi.org/10.1016/j.resuscitation.2014.08.036.

Vroom, M. B., & van Wezel, H. B. (1996). Myocardial stunning, hibernation, and ischemic preconditioning. *Journal of Cardiothoracic and Vascular Anesthesia, 10*(6), 789–799. Available from https://doi.org/10.1016/S1053-0770(96)80209-6.

Wang, C. Y., Wang, J. Y., Teng, N. C., Chao, T. T., Tsai, S. L., Chen, C. L., ... Chen, L. (2015). The secular trends in the incidence rate and outcomes of out-of-hospital cardiac arrest in Taiwan—A nationwide population-based study. *PLoS One, 10*(4), e0122675. Available from https://doi.org/10.1371/journal.pone.0122675.

Whitaker, R. M., Stallons, L. J., Kneff, J. E., Alge, J. L., Harmon, J. L., Rahn, J. J., ... Schnellmann, R. G. (2015). Urinary mitochondrial DNA is a biomarker of mitochondrial disruption and renal dysfunction in acute kidney injury. *Kidney International, 88* (6), 1336–1344. Available from https://doi.org/10.1038/ki.2015.240.

Yue, R., Xia, X., Jiang, J., Yang, D., Han, Y., Chen, X., ... Zeng, C. (2015). Mitochondrial DNA oxidative damage contributes to cardiomyocyte ischemia/reperfusion-injury in rats: Cardioprotective role of lycopene. *Journal of Cell Physiology, 230* (9), 2128–2141. Available from https://doi.org/10.1002/jcp.24941.

Zhang, M., Michael, L. H., Grosjean, S. A., Kelly, R. A., Carroll, M. C., & Entman, M. L. (2006). The role of natural IgM in myocardial ischemia-reperfusion injury.

Journal of Molecular and Cellular Cardiology, *41*(1), 62–67. Available from https://doi.org/10.1016/j.yjmcc.2006.02.006.

Zhang, Q., Raoof, M., Chen, Y., Sumi, Y., Sursal, T., Junger, W., ... Hauser, C. J. (2010). Circulating mitochondrial DAMPs cause inflammatory responses to injury. *Nature*, *464*(7285), 104–107. Available from https://doi.org/10.1038/nature08780.

Zhao, Z. Q., Corvera, J. S., Halkos, M. E., Kerendi, F., Wang, N. P., Guyton, R. A., & Vinten-Johansen, J. (2003). Inhibition of myocardial injury by ischemic postconditioning during reperfusion: Comparison with ischemic preconditioning. *American Journal of Physiology- Heart and Circulatory Physiology*, *285*(2), H579–H588. Available from https://doi.org/10.1152/ajpheart.01064.2002.

CHAPTER 4

Cardioprotection

Abstract

Cardioprotection is a very challenging area in the field of cardiovascular science. Myocardial damage accounts for up to 50% of injury due to reperfusion and yet there is no effective strategy to prevent this cardiac damage to reduce the burden of heart failure. In this chapter we discuss briefly cardioprotection and its strategies.

Keywords: Cardioprotection; oxidative stress; inflammation and apoptosis

Contents

Cardiovascular diseases (CVDs) are the most common cause of mortality worldwide and an estimated 17.9 million people have died due to CVDs in 2016, that is, one-third of global deaths (World Health Organization, 2016). There are about 2.19 million deaths in Europe every year due to ischemic heart diseases (World Health Federation, 2016). In Europe, Italy is performing the second highest number of cardiac procedures, over 100,000 in 2014 (Eurostat, 2016). The last century has seen major advances in cardiovascular science from noninvasive procedures to interventional, coronary artery bypass graft surgery and cardiac transplantation. In spite of all these major advances, cardiovascular diseases are still a growing cause of morbidity and mortality in developing as well as in developed nations. This is a huge financial burden on the healthcare management system. Unfortunately, basic cardiovascular science has not been able to achieve cardioprotection in the era of growing cardiovascular diseases, mostly because of failure into translation

Pathophysiology of Ischemia Reperfusion Injury and Use of Fingolimod in Cardioprotection
DOI: https://doi.org/10.1016/B978-0-12-818023-5.00004-2
75

of basic experimental work into clinical settings. Currently, the hot topic of research and discussion is "myocardial ischemia—reperfusion injury" to prevent irreversible myocardial damage and to prevent ventricular dysfunction, which is the leading cause of heart failure. Indeed, recent advances like ischemic pre- and postconditioning encourage and optimize toward finding novel discoveries like ischemic preconditioning opened new venues toward finding the eventual cardioprotective mechanisms. Cardioprotection is usually defined as "all mechanisms and means that contribute to the preservation of the heart by reducing or even preventing myocardial damage."

Myocardial protection related to ischemia—reperfusion injury

In recent decades, numerous studies have shown that myocardial cells possess several coping mechanisms that aim to limit the damage of ischemia/reperfusion.

The cellular mechanisms underlying certain phenomena of myocardial cell protection from ischemia/reperfusion (as the preconditioning and postconditioning) remain to be clarified in an appropriate way, but are likely to be numerous. It remains to be proven whether pharmacological interventions can improve myocardial cellular metabolism toward a more efficient use of oxygen and energy stocks can help protect the myocardium under ischemic conditions. Also, the option of using drugs to mimic the phenomenon of preconditioning or postconditioning in *in vivo* application of ischemic episodes and the ability to prevent myocardial injury with pharmacological cytoprotection by reducing apoptosis are other fascinating possibilities still developing safety assessment.

It should also be noted that other possible treatment options proposed to prevent reperfusion injury based on physiopathological methods (including Na^+/H^+ exchanger inhibitors, regional hypothermia, antiinflammatory, and reperfusion gradual) have not been tested successfully in patients with acute myocardial infarction.

Oxidative stress

Oxidative stress is more often associated with elevated levels of reactive oxygen species (ROS) or reactive nitrogen species (RNS) at the cellular and subcellular levels (Navarro-Yepes et al., 2014). However,

ROS/RNS at the suboptimal level can act as signaling molecules in maintaining the cardiovascular function (Penna, Mancardi, Rastaldo, & Pagliaro, 2009). On the other hand, increased ROS/RNS levels can induce pathology by damaging lipids, proteins, and DNA (Wiseman & Halliwell, 1996). Thus, ROS depending on concentration, the site of production, and the overall redox equilibrium of the cell will determine its biological action (beneficial or deleterious) in the tissues. Cardiovascular pathology associated with oxidative stress is observed in several cardiac diseases like ischemia/reperfusion injury (Roche & Romero-Alvira, 1995) (Fig. 4.1).

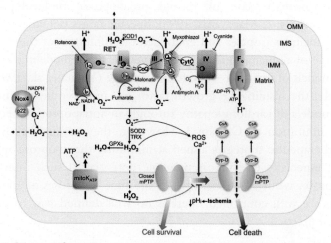

Figure 4.1 Sources of reactive oxygen species (ROS) in mitochondria and mitochondrial permeability transiting pore (mPTP) opening upon reperfusion. Mitochondrial ROS are generated by leakage of electrons from the electron transport chain, causing the incomplete reduction of oxygen to superoxide anion (O_2^-). In particular, succinate-driven RET leads to mitochondrial matrix superoxide production from complex I early during reperfusion. Mitochondrial NOX_4 also contributes to H_2O_2 generation. The mPTP formation is inhibited by acidosis and promoted by calcium and ROS. The low pH in ischemia inhibits transition pore formation but restoration of pH coupled to mitochondrial calcium overload and excessive ROS generation causes the pore to form soon after reperfusion, leading to cardiomyocyte death. Matrix CyP-D facilitates mPTP opening by enhancing its calcium sensitivity, and is the mitochondrial target of CsA, which is cardioprotective. The red lines with bar heads represent inhibition. RET is shown in dark blue. *CsA*, cyclosporin A; *CyP-D*, cyclophilin D; *GPX*, glutathione peroxidase; *IMM*, inner mitochondrial membrane; *IMS*, intermembrane space; *mitoKATP*, mitochondrial ATP-sensitive K+ channels; *OMM*, outer mitochondrial membrane; *RET*, reverse electron transport; *SOD*, superoxide dismutase; *TRX*, thioredoxin (Cadenas, 2018).

Apoptosis

Inhibition of apoptosis can limit the loss of myocardial cells induced by programmed cell death. It has been shown how different drugs have favorable effects in ischemic cardiomyopathy, including angiotensin-converting enzyme inhibitors, angiotensin II antagonists, and beta blockers, have antiapoptotic effects in animal models, through inhibition of the Renin-angiotensin system and sympathetic nervous system effectors to reduce apoptosis (Oskarsson, Coppey, Weiss, & Li, 2000). Antioxidant agents can act as antiapoptotic substances, because oxidative stress and the generation of ROS may trigger the "intrinsic" apoptosis. In a mouse model of ischemia—reperfusion, in fact, the antioxidant was able to prevent the overexpression of various proapoptotic molecules (Yaoita, Ogawa, Maehara, & Maruyama, 1998). Specific target potential to prevent apoptosis includes caspases and endonuclease. Inhibitors of these enzymes are capable of reducing infarct and left ventricular remodeling in experimental models of ischemia—reperfusion damage (Yaoita et al., 1998).

Finally, the insulin-like growth factor can improve heart function in animal models of cardiomyopathy through an antiapoptotic effect mediated by inhibition of Caspase-3. However, they cannot be over used due to therapeutic strategies aimed at inhibiting the apoptosis in clinical practice, especially due to carcinogenic potential of such intervention. Furthermore, while in animal models the time and doses of antiapoptotic drugs are well controlled, there is no well defined way to use them in clinical practice.

Postischemic mechanical assistance (PostCA)

Among the mechanical assistance systems that can be used in conditions of myocardial ischemia or infarction, and postcardiac arrest is the extracorporeal membrane oxygenator (ECMO).

It is a closed-loop system that replaces the heart and lungs of the patient when they are not able to perform, providing valid cardiopulmonary support when necessary for an extended period and allowing functional recovery of the heart and lungs. It has no therapeutic action, but allows clinicians to implement medical treatment that would exclude from the cardiac and pulmonary functionality.

There are mainly two types of ECMO, (1) veno-arterial (VA), and (2) veno-venous (Figs. 4.2 and 4.3). Both types provide respiratory

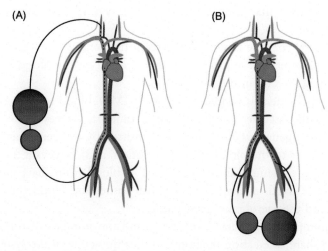

Figure 4.2 Veno-venous ECMO: two cannulation approach (A) femoral vein (for drainage) and right internal jugular for infusion,(B) both femoral veins are used for drainage and perfusion. ECMO, Extra Corporeal Membrane Oxygenation.

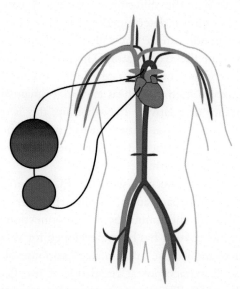

Figure 4.3 Central veno-arterial ECMO cannulation approach. ECMO, Extra Corporeal Membrane Oxygenation.

support, but only that hemodynamic support VA also provides a cardio-circulatory function.

In the first type, the blood drained from the venous access is then conveyed to the arterial access, while in the second case the blood is drained and reinfused directly in the vein.

When the patient is connected to the ECMO, blood is taken from the patient's cardiovascular system through a cannula, conveyed toward the membrane oxygenator tanks to pressure exerted by the centrifugal pump. The latter gives to the blood a circular motion, and conveys it in a vortex that, creating a vacuum allows the recall of blood from the patient, and on the other hand, a kinetic energy that pushes the blood through the spiral toward the membrane oxygenator.

The pump then generates the flow, and the number of turns determines the share of ejection fraction and thus the scope of the system. The pump types generally used are centrifugal pumps and magnetic levitation pumps. The flow produced is one of the linear types, and not pulsed.

At the level of the output line from the centrifugal pump is positioned a flow detector. The oxygenator consists of a continuous membrane, consisting of hollow fibers, in which the mixture of air and oxygen flows countercurrent to the blood, which flows perpendicularly to it; this creates a wide exchange surface between air and blood with a minimum thickness of membrane, the diffusion exchange of oxygen and carbon dioxide. The oxygenation is determined by the flow rate, while the elimination of carbon dioxide can be controlled by adjusting the speed of the flow upstream within the oxygenator. There is also a controller of the spray, which can change the flow rate of the mixture and the oxygen fraction in the mixture itself. The heat exchanger maintains a physiological body temperature or relatively hypothermic; this is necessary because the continuous passage of the entire cardiac output through the pipes, pumps, and oxygenator positioned outside the body determines a rapid heat loss resulting in hypothermia.

Before making the connection with the ECMO and during use, the patient should be on anticoagulation, usually with unfractionated heparin, and then proceeds with the cannulation and connection to the circuit of ECMO. Heparin zed coating has been developed for various circuit components in order to reduce the inflammatory response of coagulation cascade associated with extracorporeal circulation (Crea, Gaspardone, Kaski, Davies, & Maseri, 1992). The cannulation can be performed using various techniques, surgical or percutaneous: when you need a cardiopulmonary resuscitation, straws can be positioned with the percutaneous Seldinger technique to accelerate the procedure. Otherwise, surgical isolation of the vessels can be done. Cannulation can also be central or peripheral (Beckmann et al., 2011).

Cannulation can be done for central or peripheral blood that can be made to the common femoral artery, axillary artery, or common carotid artery, while the venous cannula is placed at the level of the common femoral vein or internal jugular vein; subsequently, the correctly positioned cannula is fixed with wire stitches to the patient's skin (Ailawadi & Zacour, 2009).

Specifically, in VV ECMO, the venous cannula is usually placed to the left or right common femoral vein to drain and the right internal jugular vein for the infusion. The ECMO AA, a venous cannula is placed in the inferior vena cava or right atrium, and an arterial cannula is positioned in the right femoral artery.

Occasionally the femoral vessels may not be available to cannulate the ECMO; in those cases, we can use the left common carotid artery or subclavian artery (Navia, Atik, Beyer, & Ruda Vega, 2005).

After cannulation, the patient is connected to the ECMO circuit, and the blood flow is increased to obtain satisfactory hemodynamic and respiratory parameters.

On examination, if the cardiac function of a postcardiac arrest patient is reduced minimally or not assessable, ECMO VA can provide a valid cardiopulmonary support; in these cases the device is used until the patient can recover independently or otherwise prior to implantation of a ventricular assist device, or LVAD can be used as a bridge therapy to cardiac transplantation.

The indications for the use of ECMO, other than cardiac arrest is the heart failure secondary to myocardial infarction, the refractory cardiogenic shock, the impossibility to weaning from cardiopulmonary bypass following cardiac surgery, and as an aid in conventional cardiopulmonary resuscitation extracorporeal cardiopulmonary resuscitation (Schmid, Philipp, Mueller, & Hilker, 2009).

In two observational studies, the use of ECMO in patients with cardiac arrest was associated with increased survival compared to single conventional cardiopulmonary resuscitation (Chen et al., 2008; Shin et al., 2011).

The only absolute contraindication to the use of ECMO is a preexisting condition of severe neurological damage or impossible recoveries, such as a severe neurological impairment or end-stage cancer. Relative contraindications, which must be assessed on a case-by-case basis, including uncontrollable bleeding and patients with a poor prognosis need to be extra careful while considering ECMO due to its own complications.

The main complications related to the use of a ECMO system includes bleeding, the most common complication, due to the extended period of coagulation; thromboembolic complications are infrequent but can be devastating, especially in ECMO VA. These complications can be prevented by maintaining adequate anticoagulation and observing the circuit looking for signs of a blood clot causing neurological damage, heparin-induced thrombocytopenia, and platelet transfusion becomes necessary, systemic inflammation, multiorgan dysfunction.

The consumption of coagulation factors and thrombocytopenia often requires blood transfusions, which can cause adverse reactions and secondary damage to the lungs.

Systemic inflammatory cascade contributes to various clinical outcomes, including kidney, heart, and lung damage (Wan, LeClerc, & Vincent, 1997); among the main mechanisms include surgical trauma, hemodilution, endothelial damage induced by edema, damage from ischemia—reperfusion injury of various organs, and contact activation of blood components during extracorporeal circulation (Biglioli et al., 2003). In determining the leukocyte activation and systemic inflammation, it was recently demonstrated in an experimental model of extracorporeal circulation that the oxygenator has role in leukocyte activation (Rungatscher et al., 2015). Due to these numerous side effects on hemodynamic support with ECMO can be maintained only for short periods and should be removed as soon as there is a recovery of cardiac and respiratory function or after applying the necessary treatment.

Pharmacological targets for known ischemia—reperfusion injury mechanisms

The pharmacological cardioprotective strategy to prevent acute global ischemia/reperfusion injury has been tested using different approaches. In recent years, multiple pharmacological agents including volatile anesthetic agents (Kato & Foex, 2002; Symons & Myles, 2006; Yu & Beattie, 2006), sodium hydrogen exchange inhibitors (Avkiran & Marber, 2002) and statins (Pan et al., 2004), pharmacological preconditioning (Belhomme et al., 2000; Lee, LaFaro, & Reed, 1995), and antiinflammatory strategies have been explored as potential cardioprotective therapies. However, the majority of preclinical strategies showing cardioprotective effects did not work in clinical settings (Jones et al., 2015).

Similarly, in this text, sphingosine 1-phosphate receptors were studied because of their well-established effect on inflammation, apoptosis, and oxidative stress.

References

Ailawadi, G., & Zacour, R. K. (2009). Cardiopulmonary bypass/extracorporeal membrane oxygenation/left heart bypass: Indications, techniques, and complications. *Surgical Clinics of North America*, *89*(4), 781−796. Available from https://doi.org/10.1016/j. suc.2009.05.006, vii−viii.

Avkiran, M., & Marber, M. S. (2002). Na(+)/H(+) exchange inhibitors for cardioprotective therapy: Progress, problems and prospects. *Journal of the American College of Cardiology*, *39*(5), 747−753.

Beckmann, A., Benk, C., Beyersdorf, F., Haimerl, G., Merkle, F., Mestres, C., ... Wahba, A. (2011). Position article for the use of extracorporeal life support in adult patients. *European Journal of Cardiothoracic Surgery*, *40*(3), 676−680. Available from https://doi.org/10.1016/j.ejcts.2011.05.011.

Belhomme, D., Peynet, J., Florens, E., Tibourtine, O., Kitakaze, M., & Menasche, P. (2000). Is adenosine preconditioning truly cardioprotective in coronary artery bypass surgery? *The Annals of Thoracic Surgery*, *70*(2), 590−594.

Biglioli, P., Cannata, A., Alamanni, F., Naliato, M., Porqueddu, M., Zanobini, M., ... Parolari, A. (2003). Biological effects of off-pump vs. on-pump coronary artery surgery: Focus on inflammation, hemostasis and oxidative stress. *European Journal of Cardiothoracic Surgery*, *24*(2), 260−269.

Cadenas, S. (2018). ROS and redox signaling in myocardial ischemia-reperfusion injury and cardioprotection. *Free Radical Biology and Medicine*, *117*, 76−89. Available from https://doi.org/10.1016/j.freeradbiomed.2018.01.024.

Chen, Y. S., Lin, J. W., Yu, H. Y., Ko, W. J., Jerng, J. S., Chang, W. T., ... Lin, F. Y. (2008). Cardiopulmonary resuscitation with assisted extracorporeal life-support versus conventional cardiopulmonary resuscitation in adults with in-hospital cardiac arrest: An observational study and propensity analysis. *The Lancet*, *372*(9638), 554−561. Available from https://doi.org/10.1016/s0140-6736(08)60958-7.

Crea, F., Gaspardone, A., Kaski, J. C., Davies, G., & Maseri, A. (1992). Relation between stimulation site of cardiac afferent nerves by adenosine and distribution of cardiac pain: Results of a study in patients with stable angina. *Journal of the American College of Cardiology*, *20*(7), 1498−1502.

Jones, S. P., Tang, X.-L., Guo, Y., Steenbergen, C., Lefer, D. J., Kukreja, R. C., ... Bolli, R. (2015). The NHLBI-Sponsored Consortium for preclinicAl assESsment of cARdioprotective Therapies (CAESAR): A new paradigm for rigorous, accurate, and reproducible evaluation of putative infarct-sparing interventions in mice, rabbits, and pigs. *Circulation Research*, *116*(4), 572−586. Available from https://doi.org/10.1161/CIRCRESAHA.116.305462.

Kato, R., & Foex, P. (2002). Myocardial protection by anesthetic agents against ischemia-reperfusion injury: An update for anesthesiologists. *Canadian Journal of Anaesthesia, 49* (8), 777−791. Available from https://doi.org/10.1007/bf03017409.

Lee, H. T., LaFaro, R. J., & Reed, G. E. (1995). Pretreatment of human myocardium with adenosine during open heart surgery. *Journal of Cardiac Surgery, 10*(6), 665−676.

Navarro-Yepes, J., Burns, M., Anandhan, A., Khalimonchuk, O., del Razo, L. M., Quintanilla-Vega, B., & Franco, R. (2014). Oxidative stress, redox signaling, and autophagy: Cell death versus survival. *Antioxidants & Redox Signaling, 21*(1), 66−85. Available from https://doi.org/10.1089/ars.2014.5837.

Navia, J. L., Atik, F. A., Beyer, E. A., & Ruda Vega, P. (2005). Extracorporeal membrane oxygenation with right axillary artery perfusion. *The Annals of Thoracic Surgery, 79*(6), 2163−2165. Available from https://doi.org/10.1016/j.athoracsur.2004.01.031.

Oskarsson, H. J., Coppey, L., Weiss, R. M., & Li, W. G. (2000). Antioxidants attenuate myocyte apoptosis in the remote non-infarcted myocardium following large myocardial infarction. *Cardiovascular Research, 45*(3), 679−687.

Pan, W., Pintar, T., Anton, J., Lee, V. V., Vaughn, W. K., & Collard, C. D. (2004). Statins are associated with a reduced incidence of perioperative mortality after coronary artery bypass graft surgery. *Circulation, 110*(11 Suppl 1), II45−II49. Available from https://doi.org/10.1161/01.cir.0000138316.24048.08.

Penna, C., Mancardi, D., Rastaldo, R., & Pagliaro, P. (2009). Cardioprotection: A radical view free radicals in pre and postconditioning. *Biochimica et Biophysica Acta, 1787*(7), 781−793. Available from https://doi.org/10.1016/j.bbabio.2009.02.008.

Roche, E., & Romero-Alvira, D. (1995). Role of oxidative stress in gene expression: Myocardial and cerebral ischemia, cancer and other diseases. *Medicina Clinica (Barc), 104*(12), 468−476.

Rungatscher, A., Tessari, M., Stranieri, C., Solani, E., Linardi, D., Milani, E., . . . Faggian, G. (2015). Oxygenator is the main responsible forlLeukocyte activation in experimental model of extracorporeal circulation: A cautionarytTale. *Mediators of Inflammation, 2015*, 484979. Available from https://doi.org/10.1155/2015/484979.

Schmid, C., Philipp, A., Mueller, T., & Hilker, M. (2009). Extracorporeal life support - systems, indications, and limitations. *The Thoracic and Cardiovascular Surgery, 57*(8), 449−454. Available from https://doi.org/10.1055/s-0029-1186149.

Shin, T. G., Choi, J. H., Jo, I. J., Sim, M. S., Song, H. G., Jeong, Y. K., . . . Lee, Y. T. (2011). Extracorporeal cardiopulmonary resuscitation in patients with inhospital cardiac arrest: A comparison with conventional cardiopulmonary resuscitation. *Critical Care Medicine, 39*(1), 1−7. Available from https://doi.org/10.1097/CCM.0b013e3181feb339.

Symons, J. A., & Myles, P. S. (2006). Myocardial protection with volatile anaesthetic agents during coronary artery bypass surgery: A meta-analysis. *British Journal of Anaesthesia, 97*(2), 127−136. Available from https://doi.org/10.1093/bja/ael149.

Wan, S., LeClerc, J. L., & Vincent, J. L. (1997). Inflammatory response to cardiopulmonary bypass: Mechanisms involved and possible therapeutic strategies. *The Chest, 112* (3), 676−692.

Wiseman, H., & Halliwell, B. (1996). Damage to DNA by reactive oxygen and nitrogen species: Role in inflammatory disease and progression to cancer. *The Biochemical Journal, 313*(Pt 1), 17−29.

World Health Organization, 2016. https://www.who.int/news-room/fact-sheets/detail/cardiovascular-diseases-(cvds).

Yaoita, H., Ogawa, K., Maehara, K., & Maruyama, Y. (1998). Attenuation of ischemia/reperfusion injury in rats by a caspase inhibitor. *Circulation*, *97*(3), 276—281.

Yu, C. H., & Beattie, W. S. (2006). The effects of volatile anesthetics on cardiac ischemic complications and mortality in CABG: A meta-analysis. *Canadian Journal of Anaesthesia*, *53*(9), 906—918. Available from https://doi.org/10.1007/bf03022834.

Further reading

Chazov, E. I., Matveeva, L. S., Mazaev, A. V., Sargin, K. E., Sadovskaia, G. V., & Ruda, M. I. (1976). [Intracoronary administration of fibrinolysin in acute myocardial infarct]. *Terapevticheskii Arkhiv*, *48*(4), 8—19.

Effectiveness of intravenous thrombolytic treatment in acute myocardial infarction. Gruppo Italiano per lo Studio della Streptochinasi nell'Infarto Miocardico (GISSI). (1986). *The Lancet*, *1*(8478), 397—402. https://www.ncbi.nlm.nih.gov/pubmed/2868337.

Ginks, W. R., Sybers, H. D., Maroko, P. R., Covell, J. W., Sobel, B. E., & Ross, J., Jr. (1972). Coronary artery reperfusion. II. Reduction of myocardial infarct size at 1 week after the coronary occlusion. *Journal of Clinical Investigation*, *51*(10), 2717—2723. Available from https://doi.org/10.1172/jci107091.

Maroko, P. R., Libby, P., Ginks, W. R., Bloor, C. M., Shell, W. E., Sobel, B. E., & Ross, J., Jr. (1972). Coronary artery reperfusion. I. Early effects on local myocardial function and the extent of myocardial necrosis. *Journal of Clinical Investigation*, *51*(10), 2710—2716. Available from https://doi.org/10.1172/jci107090.

Menees, D. S., Peterson, E. D., Wang, Y., Curtis, J. P., Messenger, J. C., Rumsfeld, J. S., & Gurm, H. S. (2013). Door-to-balloon time and mortality among patients undergoing primary PCI. *New England Journal of Medicine*, *369*(10), 901—909. Available from https://doi.org/10.1056/NEJMoa1208200.

Reimer, K. A., & Jennings, R. B. (1979). The "wavefront phenomenon" of myocardial ischemic cell death. II. Transmural progression of necrosis within the framework of ischemic bed size (myocardium at risk) and collateral flow. *Laboratory Investigation*, *40*(6), 633—644.

Reimer, K. A., Lowe, J. E., Rasmussen, M. M., & Jennings, R. B. (1977). The wavefront phenomenon of ischemic cell death. 1. Myocardial infarct size vs duration of coronary occlusion in dogs. *Circulation*, *56*(5), 786—794.

Rentrop, K. P., Blanke, H., Karsch, K. R., Wiegand, V., Kostering, H., Oster, H., & Leitz, K. (1979). Acute myocardial infarction: Intracoronary application of nitroglycerin and streptokinase. *Clinical Cardiology*, *2*(5), 354—363.

Van de Werf, F. (2014). The history of coronary reperfusion. *European Heart Journal*, *35*(37), 2510—2515. Available from https://doi.org/10.1093/eurheartj/ehu268.

CHAPTER 5

Experimental models for ischemia—reperfusion injury

Abstract

This chapter consists of clinical experimental models such as in-vivo rat model of global myocardial ischemia—reperfusion injury including sudden cardiac arrest, cardioplegic arrest related to cardiopulmonary bypass surgery, and heterotopic cardiac transplantation validated models. These experimental models have potential to reflect real cardiovascular diseases with ischemia—reperfusion injury.

Keywords: Myocardial ischemia—reperfusion; cardiac arrest; cardioplegic arrest; cardiopulmonary bypass surgery; heterotopic cardiac transplantation

Contents

Introduction

This chapter covers clinical experimental models that have the potential to reflect real cardiovascular diseases and procedures involving ischemia—reperfusion injury.

Cardiac arrest by inducing ventricular fibrillation

Sudden cardiac arrest is a major cause of mortality today (Nolan, et al., 2008) and postresuscitation mortality remains more than 50% (Stub, Bernard, Duffy, & Kaye, 2011). Cardiac arrest (CA) and cardiopulmonary resuscitation both are associated with global myocardial ischemia—reperfusion injury that induce myocardial dysfunction, leading to poor prognosis and adverse outcomes (Mongardon, Dumas, Ricome, Grimaldi, Hissem,

Pathophysiology of Ischemia Reperfusion Injury and Use of Fingolimod in Cardioprotection
DOI: https://doi.org/10.1016/B978-0-12-818023-5.00015-7

Pene, & Cariou, 2011). Currently, Extracorporeal Life Support is an effective way to treat cardiogenic shock or CA because of its great potential to provide quick circulatory support via peripheral vascular access (Nichol, Karmy-Jones, Salerno, Cantore, & Becker, 2006).

In this field of cardiovascular science, developing cardiac arrest (CA) models that encompass most cardiovascular variables observed in routine human cases and that may play additional important role to study different effects of CA in heart and brain. Presently, many models are available to induce CA in rats, including CA induction by a rapid intraatrial injection of potassium chloride, delivering current to the right ventricular endocardium; transesophageal cardiac pacing, transthoracic electrical fibrillation; chest compression; compression of the heart vessels against the chest wall by use of microsurgical instruments; asphyxia; and simultaneous aortic occlusion and right atrial occlusion created by an arterial and venous balloon catheter, respectively. All of these models have a few advantages and disadvantages. Although the ventricular fibrillation—induced model mimics the "square wave" type of myocardial insult (rapid loss of pulse and pressure) commonly seen in adult humans at the onset of CA, here we will use a modified, simple, and reliable VF technique induced in rats as a model of CA that could be useful in studying the mechanisms of myocardial ischemia-induced injury and the effects of fingolimod on it.

Anesthesia induction

Rats were anesthetized using 5% isoflurane in 50% O_2 in a 5-L plastic induction box. After induction, orotracheal intubation was done with a 14 G cannula, and the rats were ventilated by a mechanical device for small animals (Harvard Model 687, Harvard Apparatus, Holliston, MA). The tidal volume was 6 mL/kg and the breathing rate was 50—60 breaths/minutes with an air and oxygen mixture/fraction of inspired oxygen (FiO$_2$) = 0.5. Arterial blood gases were assessed to adjust ventilation to maintain arterial CO_2 tension (PaCO$_2$) of 35—45 mmHg. Anesthesia was maintained with 2.5% isoflurane, and intraperitoneally pancuronium-bromide (2 mg/kg) was used to maintain muscle relaxation. Rats were secured in supine position on a heating board. Throughout the experiment, electrocardiogram (ECG) was monitored by using limb leads. The right femoral artery was isolated and a miniaturized catheter of 2-Fr diameter (model SPR 838, Millar Instruments, Houston, TX) was inserted for monitoring of systemic blood pressure. To administer treatment, femoral vein cannulation was done with 24 G cannula.

Experimental model for cardiac arrest and extracorporeal resuscitation in rat

Access to the heart was achieved through a median sternotomy followed by the opening of the chest with a retractor. In order to speed up and simplify the procedures after the cardiac arrest, prior preparation of extracorporeal device cannulation need to be done. After recording, the skin and the subcutaneous planes in the midline of the neck were identified as the trachea and immediately to the right of neurovascular bundle, a venous cannula (a modified version with four holes of a caliber of 5 French catheter) was advanced through the right external jugular vein to the right atrium allowing an excellent venous drainage. The left common carotid artery was cannulated with a 24 G catheter, advanced to the aortic arch and connected to the arterial perfusion of the circuit as shown in Fig. 5.1. Full heparinization (heparin 500 IU/kg) was performed immediately before the start of extracorporeal circulation. After opening the pericardium induction of cardiac arrest proceeded, which was obtained by ventricular fibrillation using a defibrillator disbursing a 3.5 mA current at

Step-by-step procedure

Induction of anesthesia and endotracheal intubation
↓
Femoral vein and artery cannulation
↓
Electrode placement at the entrance to the superior vena cava
↓
Vecuronium injection and start of mechanical ventilation
↓
Stabilization of physiological parameters
↓
Delivery of fibrillating current and cessation of ventilation
↓
Restart ventilation and initiation of resuscitation
↓
Delivery of defibrillating current
↓
Maintaining head and body temperatures at 37°C for 1 h

Figure 5.1 Schematic diagram of procedural steps for induction of VF-induced cardiac arrest. *Adapted from Dave, K. R., Della-Morte, D., Saul, I., Prado, R., & Perez-Pinzon, M. A. (2013). Ventricular fibrillation-induced cardiac arrest in the rat as a model of global cerebral ischemia.* Translational Stroke Research, 4(5). doi: 10.1007/s12975-12013-10267-12970.

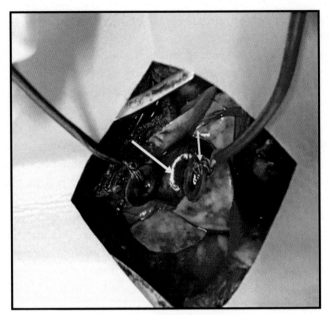

Figure 5.2 Electrical induction of VF-induced CA model. *Yellow arrow* indicates source for electrically induced VF, and *white arrow* indicates left ventricle.

60 Hz released at the level of the left ventricular epicardial region (Fig. 5.2). The current flow was maintained for 3 minutes to prevent spontaneous depolarization and cardiac arrest was maintained for 10 minutes, after which reperfusion occurred through extracorporeal circulation. Mechanical ventilation was stopped during induction of VF.

In vivo experimental rat model of cardioplegic arrest
Animal preparation
After preanaesthesia with vapors of diethyl ether, rats were orotracheal intubated with an atraumatic tube composed of a venous cannula of 14 G. The rats were then ventilated with a mechanical respirator for rodents (Inhale, Harvard Apparatus, Holliston, MA) with a mixture of oxygen and anesthetic Sevorane 2% (Abbot Laboratories, Queenborough, United Kingdom) that guaranteed anesthesia for the duration of the procedure, with a FiO$_2$ of 90%, adjusted a tidal volume of 10 mL/kg, and a frequency of 70 breaths per minute, and 0.1 mg/kg of vecuronium bromide was administered to obtain complete muscular relaxation and repeated if needed. During the surgical procedure, room temperature was

maintained between 23°C and 25°C and a layer of isolating material (cork) was placed between the animal and the operating table.

Rats were placed in supine position, and the thoracic area, area on ventral surface of the neck and hind legs were shaved, and skin was disinfected with chlorhexidine. A thermocouple microprobe was inserted into the rectum to monitor animal temperature during the experiment. ECG electrodes were attached on both front limbs and left hind limb. The right femoral artery was isolated, and a miniaturized catheter with a diameter of 2-Fr (model SPR 838, Millar Instruments, Houston, TX) for continuous monitoring of systemic blood pressure was inserted. Subsequently, the femoral vein was also cannulated with a 24-G cannula (Delta Med S.p.A., Viadana, MN, Italy), followed by administration of 500 UI/kg heparin to ensure patency and to be ready for cardiopulmonary bypass (CPB).

Surgical procedure

Access to the heart was achieved through a median sternotomy; the chest was kept open with a retractor. After incision of skin and subcutaneous planes in the midline of the neck, the trachea was identified. A venous cannula (a modified four-hole catheter with caliber 5-Fr) was advanced through the external jugular vein to the right atrium allowing excellent venous drainage. The left common carotid artery was cannulated with a 24 G (Delta Med S.p.A., Viadana, MN, Italy) catheter that was advanced to the aortic arch and connected to the line of arterial perfusion circuit. Also, 0.2 mg pancuronium was administered before CPB. Muscle relaxation was used to prevent spontaneous ventilation that often interferes with the venous return due to movement of the mediastinal structures about the venous outflow cannula (Fig. 5.3).

Perfusion circuit

The extracorporeal circulation circuit was comprised of a roller pump (Stockert SIII, Sorin, Germany), a hollow-fiber oxygenator (Sorin, Mirandola, MO, Italy), and a venous reservoir connected to a vacuum with a pressure regulator -30 mmH$_2$O to facilitate venous drainage, all connected by plastic tubing with 1.6 mm inside diameter. The total volume of filling of the extracorporeal circuit including the oxygenator was 6 mL and constituted by colloid solution and Ringer's lactate solution. The exchange surface of the gas was 450 cm^2 and the heat-exchange surface area of 15.8 cm^2. Once venous and arterial accesses were prepared, rat was connected to the circuit of CPB and maintained with a flow rate

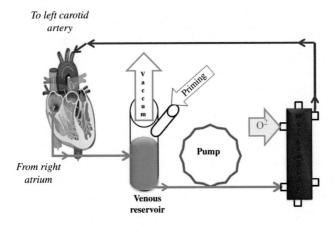

To left carotid artery

From right atrium

Venous reservoir

Pump

Maximum blood flow velocity: 100 mL/kg/min
Priming volume: 6–10 mL
Surface area of oxygenator: 450 cm^2

Figure 5.3 Schematic view of extracorporeal life support.

of 80–100 mL/kg/min and a range of mean arterial pressure of 70–90 mmHg. After ensuring adequate venous drainage and appropriate hemodynamic stability mechanical ventilation was suspended.

Cardioplegic arrest

After 10 minutes of CPB, the ascending aorta was clamped with a vascular clamp and cardiac arrest was induced by administration of 2 mL cardioplegia (St. Thomas solution). The CPB flow rate was adjusted as needed to maintain a constant venous reservoir blood level. After 10 minutes of cardiac cardioplegic arrest the aortic clamp was removed, and after a few minutes the heart usually started beating spontaneously; CPB was maintained for another 60 minutes. During the entire procedure, the mean arterial pressure remained higher than 45 mmHg. Subsequently, this was suspended, and weaning from CPB was completed with reinfusion of the blood in the circuit before rat sacrifice and removal of brain and heart (Figs. 5.4 and 5.5).

Heterotopic cardiac transplantation model of rats

A technique for a heart transplant in small animals was first described by Abbott et al. (1965) and Abbott et al. (1964), and then by Ono and Lindsey who made some changes (Ono & Lindsey, 1969; Ono, Lindsey, & Creech, 1969). Apart from these few have described the surgical techniques and

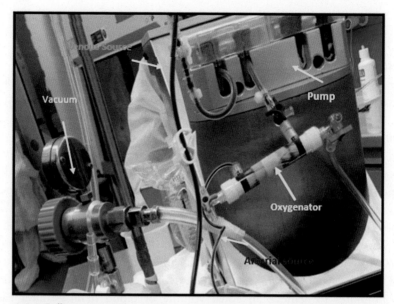

Figure 5.4 Roller pump and oxygenator during extracorporeal circulation.

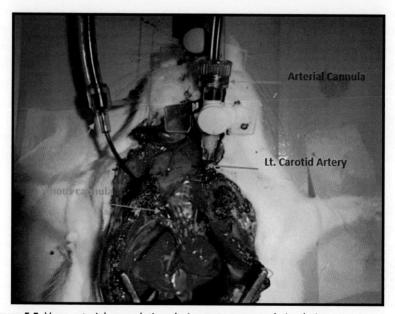

Figure 5.5 Veno-arterial cannulation during extracorporeal circulation.

anesthetic management of heterotopic cardiac transplantation in rodents (Demirsoy, Arbatli, Korkut, Yagan, & Sonmez, 2003).

Description of model

For each experiment planned two male rats weighing about 350−400 g were used. One rat was used as a receiver and the other as a donor.

Animal preparation

The donor's rat chest was opened through an incision with Mayo curved scissors to prevent injury to the major blood vessels and lungs. The cardiac exposure was done by removal of the pericardial sac. The heart was perfused with St. Thomas at 4°C temperature, with the composition of St. Thomas 9 mL and of 1 mL heparin sodium, 1000 U/mL. A 6−0 proline tie was used around the superior vena cava (SVC) and inferior vena cava (IVC). Heart perfusion was obtained by inserting a 24-G needle into the IVC. Five serial injections of 2 mL volume were delivered within 4−5 minutes to achieve an equal volume of blood drainage after each 2 mL infusion. After infusing the first 2-mL injection, the SVC was tied with 6−0 proline. Later, after infusion of the remaining solution, IVC was also tied. Perfusion is recommended at the rate of slow speed to prevent coronary vascular damage. The transverse sinus was found under the pulmonary artery and ascending aorta with 45-degree curved forceps carefully to avoid injury to the left atrium. The pulmonary artery and ascending aorta were cut together by using 60-degree sharp-angled scissors at least 4−5 mm above the origin. The SVC was dissected over the knot, and the IVC was dissected below the knot. The heart was lifted with a 4−0 proline suture placed under it, then moved downward, and also the suture was tied around the pulmonary veins. The donor's heart was preserved in St. Thomas solution at 4°C on ice during preparation of the recipient rat.

Graft implantation

The animal was positioned with head toward the left side of the surgeon. The recipient rat abdomen was opened through a midline incision with Mayo curve scissors. The intestines were moved to the left side and covered with moistened gauze with normal saline. The major vessels (IVC to the right of midline and abdominal aorta left side of the midline) were dissected and separated from the surrounding tissue gently with a cotton tip, using mini forceps only when necessary. The small iliolumbar veins were exposed and tied with 6−0 silk sutures or cauterized. Vascular miniclamps with delicate teeth distally entered first, then proximally, abdominal aorta and IVC, and isolated a 1.2-cm segment of the vessels between the jaws. The aorta was incised near the distal terminals using a 24-G needle to drain the blood. Extra-fine scissors were used to perform an arteriotomy of about 4 mm, with special care not to damage the posterior wall. The lumen was gently washed with a heparinized St. Thomas solution. The donor's heart was placed in the field in preparation for end-to-side anastomosis. The organ was positioned with the apex toward the tail and the aorta above the pulmonary artery. The aortic anastomosis was performed by inserting a point of anchoring with the proximal end of the arteriotomy (out-in) the abdominal aorta, and then a point on the aorta graft-out, and was set with a triple knot. A second (out-in) of the suture anchor has been placed on the opposite side of the first point-out the distal end of the arteriotomy. After positioning the second suture anchor, a continuous suture (5—6 points) was performed on the front wall from the proximal end toward the distal end of the arteriotomy. The heart graft was repositioned by turning it over to expose the back wall. The anastomosis was completed with a continuous running suture. The graft was turned back to expose the anterior wall of the graft where the pulmonary artery had collapsed over the aorta. An opening 5—7 mm in length was made in the receiving IVC with scissors, and the lumen was flushed with the heparinized St. Thomas solution to remove any thrombus. The IVC opening should be more (5—6 mm in length) than the aortic opening to anastomose the pulmonary artery appropriately. An out-in stitch was kept on the distal part of the pulmonary artery and in-out at the distal part of the IVC opening. A continuous suturing was performed first a on posterior wall and then at the anterior part of pulmonary artery—IVC. Before releasing micro vessel clamps, the suture line has been thoroughly checked for any expected leakage point. First, the distal clamp was released, and then after 30 seconds the proximal clamp. Anastomosis was checked throughout suture point, and if there was some bleeding point, it was controlled using cotton tips and oxidized cellulose (Tabotamp, Ethicon, and Neuchatel, Switzerland). Within a few seconds after reperfusion, spontaneous heart contractions were observed in the grafted heart. The total ischemia time in a successful operation ranged from 40 to 60 minutes. The abdominal structures were replaced with care to avoid torsion of intestines. The incision was closed with a nonabsorbable continuous suture (4—0 proline) that included the fascia and muscles and separate suturing was done for skin. The rat was kept under the heat lamp for

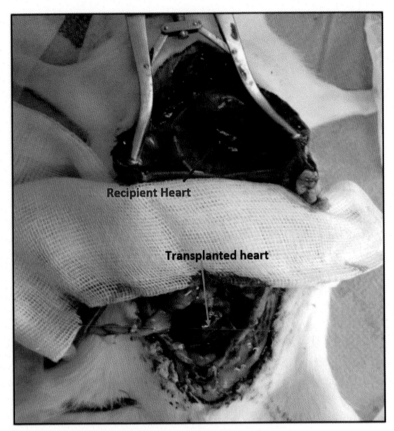

Figure 5.6 Image showing recipient and transplanted heart.

15 minutes to achieve normothermia. Rats were also kept in a warm and oxygen-rich room until sacrifice or other measurements.

Removing the weight, a constant ligation pressure could be applied and easily released. Leads were implanted subcutaneously in a lead II ECG configuration. Animals were then positioned in a specially designed Plexiglas cradle fitted to the occlusion device. Rats stabilized for 15–20 minutes before arrest. Ischemia was visually confirmed by regional cyanosis downstream of the occlusion or changes in the ECG. The heterotopic transplanted heart in vivo rat model is shown in Fig. 5.6.

Myocardial infarction and revascularization

Experimental protocol to study myocardial ischemia-reperfusion injury and hemodynamic studies on rats of all the experimental groups need to be anesthetized intraperitoneally with pentobarbitone sodium (60 mg/kg). Atropine can be administered with the anesthetic agents to keep the heart rate maintained especially during the surgery protocol and to decrease bronchial secretions. The animal body temperature needs to be monitored and maintained at 37°C throughout the experimental protocol. The incision may be given on a ventral midline to open the neck, and a tracheostomy need to be performed and the experimental animal should be on ventilation with room air from a positive pressure ventilator (Harvard, United States) using compressed air at a rate of 70 strokes/min and a tidal volume of 6 mL/kg. The right carotid artery needs to be cannulated and the cannula filled with heparinized saline may be connected to the cardiac output monitor CARDIOSYS CO-101 (Experimetria, Hungary) via a pressure transducer for measurement of mean arterial pressure(MAP) and heart rate (HR). The left jugular vein was cannulated with polyethylene tube for continuous infusion of 0.9% saline solution.

To see the cardioprotective effects of an experimental agent/drug, a thoracotomy needs to performed at the fifth intercostal space and the pericardium opened to expose the heart. The left anterior descending (LAD) coronary artery was ligated 4—5 mm from its origin by a 6—0 silk suture with a needle with atraumatic ends and ends of this ligature passed through a small vinyl tube to form a snare. After completion of the surgical procedure, the heart may be returned to its normal position in the thorax. The thoracic cavity need to be covered with saline-soaked gauze to prevent the heart from drying. The animals was monitored then allowed to stabilize for 15 minutes before LAD ligation. Myocardial ischemia was induced by one-stage occlusion of the LAD by pressing the polyethylene tubing against the ventricular wall and then fixing it in place by clamping the vinyl tube with a hemostat. A wide bore (1.5 mm) sterile metal cannula will be inserted into the cavity of the left ventricle from the posterior apical region of the heart. A bolus of heparin (30 IU) administered immediately before coronary artery occlusion for prophylaxis against thrombus formation around the snare. The animals then underwent 45 minutes of ischemia, confirmed visually in situ by the appearance of regional epicardial cyanosis and ST-segment elevation. The myocardium can be reperfused by releasing the snare gently for a period of 60 minutes. Successful reperfusion may be confirmed by visualization of arterial blood flow through the artery, appearance of hyperemia over the surface of the previously ischemia cynotic segment. At the end of the reperfusion period, animals need to be sacrificed for biochemical, immunohistochemical, and histological studies by an overdose of anesthesia and active euthanasia.

Hemodynamic monitoring and measurements

After recording, the skin and the subcutaneous planes in the midline of the neck were identified as the trachea immediately to the right of neurovascular bundle. The right carotid artery was isolated, cranially and caudally clamped. After making a small cut with microsurgical scissors a miniaturized catheter of 2-Fr diameter (model SPR 838, Millar Instruments, Houston, TX) was inserted into the carotid artery (Fig. 5.7).

The catheter was then connected to the relevant transducer and then to the unit Power-Lab (AD Instruments, Colorado Springs, CO). The latter, through the USB port, was connected to a computer for real-time display of pressure—volume curves and data logging using the Chart software (AD Instruments). With this setup, the catheter to conductance was introduced into the carotid artery and advanced to the left ventricle; the correct position is reached by following the trend of the pressure curve (Fig. 5.8).

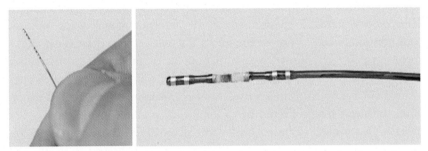

Figure 5.7 Conductance catheter for monitoring of pressures and volumes in the left ventricle (Millar Instruments, TX, United States).

(A)   (B)

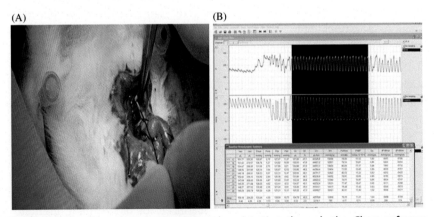

Figure 5.8 Carotid catheterization (A) and registration through the Chart software that shows the transition from the aorta to the left ventricle (B).

The signal is recorded continuously with a sampling rate of 1000/s and was thus able to monitor the hemodynamic changes during the whole duration of the intervention. The parameters can be considered including left ventricular end-systolic pressure, left ventricular end-diastolic pressure, ejection fraction, stroke volume, maximum increase in systolic blood pressure $(+dp/dt)$, and maximum decrease in diastolic blood pressure $(-dP/dt)$, maximal power, adjusted preload maximal power, and tau-Weiss constant time. The calibration of the conductance system was obtained as previously described (Pal Pacher et al., 2008). Briefly, nine cylindrical holes of known diameters between 2 and 11 mm inserted into a 1 cm high block were filled with the heparinized whole blood of rats. In this calibration, the volume conductance-linear regression of the absolute volume contained in each cylinder compared with the raw signal obtained through the conductance catheter was used for the volume calibration formula.

References

Abbott, C. P., Dewitt, C. W., & Creech, O., Jr. (1965). The transplanted rat heart: Histologic and electrocardiographic changes. *Transplantation, 3*, 432−445.

Abbott, C. P., Lindsey, E. S., Creech, O., Jr., & Dewitt, C. W. (1964). A technique for heart transplantation in the rat. *Archives of Surgery, 89*, 645−652.

Dave, K. R., Della-Morte, D., Saul, I., Prado, R., & Perez-Pinzon, M. A. (2013). Ventricular fibrillation-induced cardiac arrest in the rat as a model of global cerebral ischemia. *Translational Stroke Research, 4*(5). Available from https://doi.org/10.1007/s12975-12013-10267-12970.

Demirsoy, E., Arbatli, H., Korkut, A. K., Yagan, N., & Sonmez, B. (2003). A new technique for abdominal heart transplantation in rats. *The Journal of Cardiovascular Surgery (Torino), 44*(6), 747−750.

Mongardon, N., Dumas, F., Ricome, S., Grimaldi, D., Hissem, T., Pene, F., & Cariou, A. (2011). Postcardiac arrest syndrome: From immediate resuscitation to long-term outcome. *Annals of Intensive Care, 1*(1), 45. Available from http://dx.doi.org/10.1186/2110-5820-1-45.

Nichol, G., Karmy-Jones, R., Salerno, C., Cantore, L., & Becker, L. (2006). Systematic review of percutaneous cardiopulmonary bypass for cardiac arrest or cardiogenic shock states. *Resuscitation, 70*(3), 381−394. Available from http://dx.doi.org/10.1016/j.resuscitation.2006.01.018.

Nolan, J. P., Neumar, R. W., Adrie, C., Aibiki, M., Berg, R. A., Bottiger, B. W., & Hoek, T. V. (2008). Post-cardiac arrest syndrome: epidemiology, pathophysiology, treatment, and prognostication. A Scientific Statement from the International Liaison Committee on Resuscitation; the American Heart Association Emergency Cardiovascular Care Committee; the Council on Cardiovascular Surgery and Anesthesia; the Council on Cardiopulmonary, Perioperative, and Critical Care; the

Council on Clinical Cardiology; the Council on Stroke. *Resuscitation, 79*(3), 350−379. Available from http://dx.doi.org/10.1016/j.resuscitation.2008.09.017.

Ono, K., & Lindsey, E. S. (1969). Improved technique of heart transplantation in rats. *The Journal of Thoracic and Cardiovascular Surgery, 57*(2), 225−229.

Ono, K., Lindsey, E. S., & Creech, O., Jr. (1969). Transplanted rat heart: Local graft irradiation. *Transplantation, 7*(3), 176−182.

Pacher, P., Nagayama, T., Mukhopadhyay, P., Bátkai, S., & Kass, D. A. (2008). Measurement of cardiac function using pressure-volume conductance catheter technique in mice and rats. *Nature protocols, 3*(9), 1422−1434. Available from https://doi.org/10.1038/nprot.2008.138.

Stub, D., Bernard, S., Duffy, J., & Kaye, D. M. (2011). Post cardiac arrest syndrome. *Circulation, 123*(13), 1428.

CHAPTER 6

Role of fingolimod in cardioprotection

Abstract

The fingolimod (2-amino-2-[2-(4-octyl-phenyl) ethyl]-1.3-propanediol hydrochloride) is a synthetic structural analog of sphingosine that acts as a receptor agonist of sphingosine-1-phosphate and represents a new frontier in the treatment of ischemia-reperfusion. This drug plays its role in survival by signaling pathways to attenuate myocardial injury.

Keywords: sphingosine; first sphingosine-1-phosphate; fingolimod; ischemia—reperfusion injury; cardioprotection

Contents

Sphingosine-1-phosphate

Sphingosine-1-phosphate (S1P) levels in the plasma is mainly produced by endothelial cells, from erythrocytes, platelets, and hepatocytes is a bioactive lysophospholipid deriving from sphingomyelin, ubiquitous lipid cell membranes (Spiegel & Milstien, 2003). Sphingosine is formed by the enzyme sphingosine kinase (Fig. 6.1).

Pathophysiology of Ischemia Reperfusion Injury and Use of Fingolimod in Cardioprotection
DOI: https://doi.org/10.1016/B978-0-12-818023-5.00005-4
101

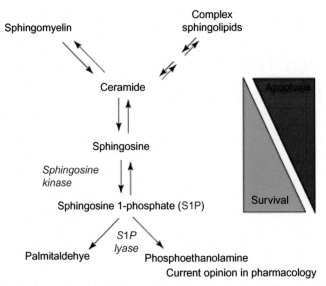

Figure 6.1 Interconversion of sphingolipids, including the formation of sphingosine-1-phosphate (S1P) from sphingosine. The effects of proapoptotic ceramide are countered by S1P, which is generally a survival signal (Kennedy, Kane, Pyne, & Pyne, 2009).

The new therapeutic perspectives of fingolimod are the result of increasing knowledge on S1P (Ahmed, Linardi, Decimo, et al., 2017; Spiegel & Milstien, 2003). Among the many effects are cytoprotective antioxidants, like immunosuppressive (Xiang et al., 2013), and its possible role in reducing ischemia–reperfusion (Karliner, 2013).

The S1P mediates different physiological functions (Groves, Kihara, & Chun, 2013), including cell proliferation, differentiation, and survival, as well as the reorganization of the cytoskeleton, the formation of cyto-plasmic extensions, cell motility and chemotaxis, intercellular adhesion, and formation of the junctions between cells; it is therefore involved in many physiological aspects of the organism, such as immunity, maintain-ing the tone and pulmonary vascular smooth muscle, endothelial barrier integrity, and both the cardiovascular system and central nervous system feature morphogenesis.

Although S1P is derived primarily from erythrocytes, other sources include the platelets, mast cells, endothelial cells, fibroblasts, and central nervous system (Fukushima, Ishii, Contos, Weiner, & Chun, 2001; Gardell, Dubin, & Chun, 2006; Schwab & Cyster, 2007; Spiegel & Milstien, 2003).

S1P performs its functions and makes a comprised up to five G-protein receptors (S1P1, S1P2, S1P3, S1P4, S1P5) (Noguchi & Chun, 2011).

Recent studies have shown that S1P reduces ischemia-reperfusion injury in the liver (Man et al., 2005), in the kidney (Delbridge, Shrestha, Raftery, El Nahas, & Haylor, 2007), and in the brain (Wacker, Park, & Gidday, 2009; Wei et al., 2011).

Note that sphingosine-1-phosphate are also able to increase the survival of cardiomyocytes during episodes of hypoxia; evidence emerged from in vitro studies (Karliner, Honbo, Summers, Gray, & Goetzl, 2001; Tao, Zhang, Vessey, Honbo, & Karliner, 2007), it can also reduce the size of the infarcted area in productions of isolated hearts ex vivo (Jin et al., 2002; Lecour et al., 2002).

Fingolimod

The drug fingolimod (2-amino-2-[2-(4-octyl-phenyl) ethyl]-1.3-propane-diol hydrochloride) is a synthetic structural analog of sphingosine that acts as a receptor agonist of S1P, and represents a new frontier in the treatment of ischemia-reperfusion.

The fingolimod hydrochloride is a powder that is freely soluble in water, alcohol, and propylene glycol. Currently, fingolimod represents a unique receptor agonist of S1P approved by the Food and Drug Administration (FDA) for clinical use in humans (Brinkmann et al., 2010) (Fig. 6.2).

History of fingolimod

The molecule of fingolimod, also known as FTY720, was synthesized for the first time in 1995; fingolimod is derived from myriocin (Chiba, 2009) (ISP-1) which is a metabolite from fungus *Isaria candidate* and *Myrothecium verrucaria* (Hsu et al., 1989), the active ingredient of natural origin for

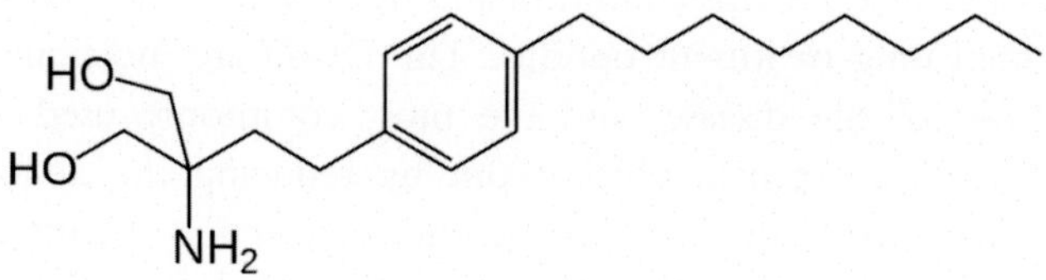

Figure 6.2 Molecular structure of fingolimod.

immunosuppressive agents found in 1992 by a group of Japanese research-ers; by this, with the proper chemical modifications, then arrived to the drug, with better tolerability.

On September 22, 2010, the FDA approved its use as a first-line ther-apy for relapsing-remitting multiple sclerosis, and on March 17, 2011, the European Commission issued an authorization making it commercially valid throughout the European Union. On November 22, 2011, fingoli-mod was approved in Italy.

Use of fingolimod in clinical settings

Fingolimod was defined by the European Medicines Agency (EMA) as a first-line therapy in patients with severe and rapidly progressing disease forms, or as second-line therapy in the case of failure with the use of interferon beta-1a (IFN-β-1a).

In European Union, fingolimod is approved for a use as monotherapy, as a disease-modifying drug (DMD), in selected patients with a form highly active relapsing-remitting multiple sclerosis.

The efficacy of fingolimod consists not only in reducing the recur-rence rates of multiple sclerosis (MS) and in improved metering suggestive of active disease made by MRI, both compared with placebo than with INF-β, but also in reducing the progression of disability compared to placebo (Cohen et al., 2010; Kappos et al., 2010).

Fingolimod is different from existing therapies available for multiple sclerosis as it primarily represents the first oral treatment in 0.5 mg capsules.

For the last two decades, considerable progress has been made to effec-tively control the huge challenge in the treatment of MS that is relapsing and remitting of the disease (Aguiar, Batista, & Pacheco, 2015).

Multiple sclerosis is a chronic inflammatory disease of the central ner-vous system, caused by abnormal functioning of the immune system, sticking myelin the substance of coating of nerve fibers. Several DMDs have been approved for this condition and new molecules, such as fingoli-mod, are increasing treatment options. The DMD are molecules that act on the causes of the disease, and the most commonly used today are immunosuppressive agents, which work by reducing the action of the immune system and thus hindering the destruction of myelin, and immune-modulating agents that alter the delicate balance of the immune system and decrease its action, with the objective of limiting the attack on myelin (Lebrun et al., 2011).

Among the DMDs, new agent fingolimod with innovative mechanism of action and its active metabolite "sphingosine 1-phosphate" resulting from the enzyme sphingosine kinase phosphorylation (Brinkmann et al., 2010), acts as a S1P receptor agonist for S1PR1, S1PR3, S1PR4, and S1PR5 (Brinkmann et al., 2002). Initially, Sphingosine 1-phosphate binding to its receptor act as a agonist, but in the long term, it leads to internalization of receptors and consequent degradation proteasome level, leading to down regulation of S1P receptors (Graler & Goetzl, 2004; Oo et al., 2007).

Fingolimod modulates the S1P receptors to induce a lymphocyte downregulation, making memory T lymphocytes, and T naive remain sequestered within lymph nodes, this mechanism has the potential to reduce the traffic of these cells "pathogenesis" in the central nervous system (Chun & Brinkmann, 2011; Chun & Hartung, 2010).

As fingolimod is highly lipophilic, it can cross the blood-brain barrier and penetrate into the central nervous system (Foster et al., 2007). During last few years, scientists found stronger evidence about direct action of S1P receptors on oligodendrocytes, astrocytes, and neurones (Cohen & Chun, 2011).

Effects of fingolimod

The safety profile of fingolimod has generally proved safe and well tolerated, presenting medium grade-moderate adverse events including sinus bradycardia, atrioventricular block, infections, increased liver enzymes, hypertension, and macular edema.

The cardiovascular effects are a major source of concern in clinical settings, especially after the administration of the first dose of the medicine (Aguiar et al., 2015).

The cardiovascular effects of fingolimod have been assessed through four clinical trials IV: *TRANSFORMS, FREE DOMS, FREEDOMS II,* and *FIRST* (Calabresi et al., 2014; Cohen et al., 2010; Gold et al., 2014; Kappos et al., 2010; Khatri et al., 2011; Wang et al., 2014). In recent studies, it has been observed that reduction in heart rate is similar in both patients with and without cardiovascular risk factors, a transient reduction and dose-dependent. The heart rate is observed after the first dose, reducing the therapy continued, returning to baseline heart rate after about a month. Associated symptoms are rare, transient, and usually without clinical consequences. A recent multicenter phase IV study called "The Evaluate Patient Out Comes" confirmed these findings.

These studies demonstrated the delay in atrioventricular conduction using fingolimod is uncommon and effect is transient, and usually without clinical consequences.

Additional features of this drug are that it does not change the duration of the QRS complex or prolong the QT interval (Schmouder et al., 2006). It also does not show significant effects on platelet counts or function.

In regards to long-term treatment, in a 4-year trial extending the FREEDOMS trial, the incidence of cardiovascular adverse events remained at the same level observed in the first 2 years of the study, indicating that long-term treatment has no impact on cardiovascular disease. This result was later confirmed in the TRANSFORMS trial, which was extended to 4.5 years.

The effects on heart rate and atrioventricular conduction may recur when treatment is resumed after more than 2 weeks of suspension.

Basic mechanisms of the cardiovascular effects of fingolimod

The cardiovascular effects caused by fingolimod are due to the presence of S1P receptors in the heart and blood vessels (Means & Brown, 2009; Peters & Alewijnse, 2007); in particular, the S1P1 receptor predominates at the sinoatrial node, atrioventricular node, cardiomyocytes and endothelial cells, whereas the S1PR2 receptor is found primarily in arterial smooth muscle cells, cells which are at the same concentrations S1PR1 and S1PR2 (Chae, Proia, & Hla, 2004; Liu et al., 2000; Salomone et al., 2003; Singer et al., 2005).

In the heart, S1P1 receptor activation causes G-protein coupled receptor activation, followed by the activation of potassium channels called "inwardly-rectifying potassium G protein-coupled channels" (Koyrakh, Roman, Brinkmann, & Wickman, 2005) potassium efflux; that follows goes to hyperpolarize the cell membrane, inhibiting depolarization and reducing the excitability and robotics (Walsh, 2011). The result is, therefore, a reduction in heart rate.

The S1P1 receptor internalization is responsible for the negative chronotropic effects of impermanence and dromotropic (respectively reducing the heart rate and speed of conduction of electrical impulses in the heart) induced by fingolimod. Both atropine, a muscarinic receptor antagonist, and isoprenaline, a beta-1 receptor agonist and beta-2, reverse the effects of fingolimod (Kovarik, Riviere, et al., 2008; Kovarik, Slade, et al., 2008).

In endothelial cells, S1P1 receptor activation leads to phosphorylation of protein kinase B and activation of endothelial nitric oxide synthase, events that increase the production of nitric oxide (Brinkmann, 2007; Rosen & Goetzl, 2005; Tolle et al., 2005).

The S1P1 receptor in the smooth muscle cells of arteries leads to the release of calcium from intracellular reserves, increasing its concentration within the cells themselves and causing the contraction of smooth muscle cells and arterial vasoconstriction. Consequently the pressure does not change significantly in the acute phase following administration of fingolimod for this balance between vasodilation effect and vasoconstriction from it.

At the level of smooth muscle, smooth muscle level, as a result of the internalization of S1PR1 (functional antagonism) and S1PR3 receptor binding is favored, which leads to the opening of calcium channels increasing intracellular concentration, causing smooth muscle cell contraction; these effects are responsible for the slight increase in blood pressure, about 1−2 mmHg, observed after 2 months of treatment with fingolimod 0.5 mg/day reaching the peak pressure at 6 months and then remained stable (Cohen et al., 2010; Kappos et al., 2010).

Reorganization of the cytoskeleton, cell geometry and stabilization of intercellular junctions are in result of activation of S1PR1 level at endothelial cell level (Waeber, Blondeau, & Salomone, 2004); these effects strengthen the endothelial barrier and diminish the permeability. In the arterial smooth muscle and endothelium receptors agonist effect is transitory due to internalization of receptors, followed by activation of S1PR3, which leads to increased permeability of the endothelial barrier for breaking the intercellular junctions.

Recommendations for safe use of fingolimod

For safe use of fingolimod, with particular reference to cardiovascular adverse events, the EMA recommends an observation period of 6 hours after the first dose of the drug in all patients, to assess the signs and symptoms of bradycardia. Heart rate and blood pressure should be measured and recorded either before administration of the first dose every hour throughout the observation period; an electrocardiogram should be performed both before the first dose, then 6 hours later. During those 6 hours continuous ECG monitoring is recommended in real time. The observation period should be extended for at least another 2 hours if the heart rate lowers at after 6 hours; in the latter case, you should monitor until heart rate does not increase.

Conditions that warrant extension of the period of observation until at least the next day, or until their resolution, include:
- heart rate less than 45 bpm to 6 hours;
- QT interval ≥ 500 ms;
- second-degree atrioventricular block of new onset in 6 hours;
- third-degree atrioventricular block, regardless of the time of onset; and
- need for drug therapy during the observation period; in this case, monitoring should be conducted throughout the night and repeated after the second dose of medication.

Administration of fingolimod is not recommended in the following cases:
1. Patients predisposed to developing cardiac arrhythmia such as those with heart block Mobitz II, third-degree heart block, sick sinus syndrome, or with a history of symptomatic bradycardia or recurrent syncope.
2. Patients such as those with non tolerant ischemia heart disease, cerebrovascular disease, heart failure, recent history of myocardial infarction or cardiac arrest.
3. Patients being treated with class Ia antiarrhythmic drugs or class III.
4. Patients taking heart rate-lowering medications like beta blockers and nondihydropyridine calcium channel blockers.

Fingolimod is not absolutely contraindicated in such cases and may be administered when the benefits outweigh the risks. Monitoring should be extended to at least the first night in the first, second, and fourth cases if rate lowering drug therapy cannot be suspended.

Since the effects of fingolimod on heart rate and atrioventricular conduction can occur at the time of resumption of the drug, depending on the duration of the suspension, it is also recommended to repeat monitoring after the first dose when therapy is interrupted for one or more days during the first 2 weeks of treatment, for more than 7 days during the third and fourth week, or for 2 or more weeks if after a month or more of therapy.

Mechanisms of cardioprotection by fingolimod

S1P on cardiomyocytes binds to G-protein-coupled receptors including S1PR1, S1PR3, and S1PR4, leading to activation of numerous signal transduction pathways involved in cardioprotective action "intracellular" potentially, S1P1, in particular, is the main receptor of S1P in cardiomyocytes, and typically activates downstream signal transduction pathways reperfusion injury salvage kinase (RISK) and surviving activating factor enhancement (SAFE) (Ahmed, Linardi, Muhammad, et al., 2017; Karliner, 2013; Waeber & Walther, 2014).

The cardioprotective effect of fingolimod are mediated by activation of reperfusion injury salvage kinase (RISK) and survivor activating factor enhancement (SAFE), because both the antiapoptotic and antioxidant actions reducing the size of the heart attack is reversed using concomitantly monoamine pathways RISK and SAFE. The main molecular cascades can inhibit mitochondrial transition pore openings are the RISK, and SAFE pathways (Heusch, 2013, 2015; Heusch, Musiolik, Gedik, & Skyschally, 2011), and previous in vitro studies claim that S1P can activate these pathways (Kelly et al., 2010; Lecour et al., 2002; Zhang et al., 2007).

The downstream pathway S1P-R is a potential therapeutic target to prevent peri-infarct. Evidence arising from previous preclinical studies suggests that S1P represents a very promising pharmacological target for mitigating the damage from myocardial ischemia-reperfusion. In the ventricular cardiomyocytes of rats, adults, and babies (Karliner et al., 2001; Tao et al., 2007) sphingosine 1-phosphate increased cardiomyocyte survival during episodes of hypoxia. S1P also induced resistance to ischemia-reperfusion injury in rat hearts in ex vivo studies of wild-type rats (Jin et al., 2002; Lecour et al., 2002). In these studies, the hearts of mice lacking the enzyme sphingosine kinase, an enzyme necessary for the synthesis of S1P, developed more ischemia-reperfusion myocardial damage compared with controls (Gomez et al., 2011; Jin et al., 2007).

To provide evidence for cardioprotective role of S1P in reducing the size of myocardial infarct, S1P receptors knockout mice showed an area of larger infarct as compared to control group (Means et al., 2007; Theilmeier et al., 2006).

The metabolism of sphingosine also seems to be a key mediator in preconditioning and postconditioning, recognized strategies for cardioprotection (Yellon & Hausenloy, 2007). Indeed, both preconditioning and postconditioning can reduce the size of the infarct, which does not take place in the hearts of sphingosine kinase or S1P receptors missing (Gomez et al., 2011; Jin et al., 2007; Lecour et al., 2002; Theilmeier et al., 2006). The mitochondrial transition pore opening represents the final step that leads to apoptosis of cardiomyocytes in ischemia-reperfusion injury and can the trigger oxidative stress that characterizes it. Therefore, by preventing the mitochondrial transition pore opening you can reduce infarct size (Piot et al., 2008; Skyschally, Schulz, & Heusch, 2010).

Previous studies have shown great benefits in immunosuppression for the prevention of ischemia-reperfusion injury (Yang et al., 2006). Fingolimod is a drug that is already in use for the purpose of treating multiple sclerosis, by exerting an immunomodulatory action able to regulate the traffic of lymphocytes from peripheral blood and tissues to lymph nodes and reduces the output of lymphocytes from lymph nodes themselves (Brinkmann et al., 2010; Spiegel & Milstien, 2003).

In a recent trial conducted by Wang et al. (2014) using a mouse model of spontaneous coronary atherosclerotic occlusion based, saw itself as in mice treated with fingolimod got a reduction in the size of the infarcted area (ex vivo) as well as reduce mortality. In these same mice there was reduced count of CD4 and CD8T cells and an increased number of T regulatory cells, suggesting how the immunosuppressive effects of fingolimod contribute to its cardioprotective properties.

Together these findings emphasize that S1P and synthetic analog fingolimod may play a key role in the prevention of ischemia-reperfusion injury (Fig. 6.3).

Evidence shows the cardioprotective effects of fingolimod. Preliminary in-vitro studies have shown that S1P receptor activation mediated by reducing oxidative stress, apoptotic effects, inhibit inflammatory mediators, and reducing the loss of cardiomyocytes in hypoxic conditions (Hofmann et al., 2009; Karliner, 2013; Zhang et al., 2007).

Studies on preparations of mouse and rat isolated hearts have shown that fingolimod is also able to reduce ischemia-reperfusion injury and improve myocardial function (Hofmann et al., 2009). Despite this evidence various cardioprotective effects, they have not yet been studied in models of large animals.

Experimental study conducted by Santos-Gallego et al. (2016) using a model of myocardial ischemia-reperfusion conducted on pigs, according to a short-term protocol (administration of fingolimod 15 minutes before reperfusion or saline in controls) and a long-term protocol (administration

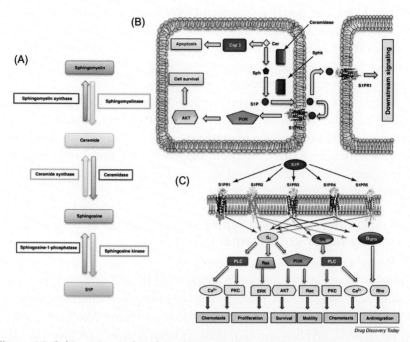

Figure 6.3 Sphingosine-1-phosphate (S1P) signaling. (A) The multistep production of S1P is governed by several enzymes, including sphingomyelinase, ceramidase, and sphingosine kinase (SphK). (B) Ceramidase converts ceramide (Cer) into sphingosine (Sph), which is later phosphorylated by SphK into S1P. Cer leads to the activation of caspase-3 (cap3), resulting in apoptosis, while S1P can also be transported to the extracellular milieu, where it acts in an autocrine or paracrine fashion by binding to the S1P receptor (S1PR) and activating phosphoinositide 3-kinase (PI3K) and AKT, leading to cell survival. (C) Binding of S1P to its receptor initiates several downstream signaling pathways via coupling to respective G-proteins. Cartoon diagrams of S1PR1−5 were generated using PyMOL. All protein structures except S1PR1 were modeled using MODELLER 9.13S1PR2, S1PR3, S1PR4, and S1PR5 protein models were generated using the template structure of the S1PR1 (Protein Data Bank id: 3V2W). Transmembrane helices of different receptors are shown in different colors to differentiate them in the membrane. *ERK*, extracellular signal-regulated kinase; *PKC*, protein kinase C; *PLC*, phospholipase C (Arish et al., 2016).

of fingolimod 15 minutes before reperfusion or saline in controls; the same treatment then once a day for 3 days). S1P receptor activation mediated by fingolimod, given before the myocardial reperfusion resulted in a significant reduction of cardiomyocyte apoptosis in the outskirts of ischemic myocardium, proven by terminal deoxynucleotidyl transferase dUTP nick end labeling (TUNEL), a technique that allows detecting fragmented DNA through the labeling of nucleic acid terminal residue. It is commonly used to detect fragmented DNA resulting from the cascade of activation of apoptosis and relies on the presence of breaks in the DNA sequence that can be made evident through the use of terminal deoxynucleotidyl transferase, an enzyme that catalyzes the addition of dUTP, which is subsequently detected by a marker.

Fingolimod plays cardioprotective role by attenuating oxidative stress leading to reduction in myocardial infarct size in pigs treated with this medication, associated with a concentration of 8-hydroxydeoxyguanosine, markers of oxidative stress; always in support of this aspect of fingolimod, antioxidant superoxide dismutase enzyme activity is increased in animals treated with fingolimod.

Activation of reperfusion injury salvage kinase and surviving activating factor enhancement pathways

To reduce cardiomyocyte apotosis, fingolimod mediates its action through activation of signal transduction pathways "RISK and SAFE". For confirmation of fingolimod results, the simultaneous administration of Wortmannin, an inhibitor of pathway RISK (including Akt/ERK/GSK-3 β), or AG490, SAFE pathway inhibitor (Janus kinase/STAT3), cancelled the anti-apoptotic effect towards the cardiomyocytes. For further proof of activation by fingolimod, pathways that regulate the protection of cardiomyocytes ischemia-reperfusion damage (Ahmed, Linardi, Muhammad, et al., 2017; Heusch, 2013, 2015), it has been shown that Akt, GSK-3, ERK1 1/β, and phosphorylated STAT3 were markedly on the edge of the ischemic myocardium in pigs treated with fingolimod 1 day after myocardial infarction, reflecting its cardioprotective effect. In contrast, Akt and ERK1/2 signaling pathways phosphorylation reduced remote myocardial infarction after 1 month, leading to prevent myocardial remodeling. Thus, the downstream signaling cascades of S1P-R mediate the activation of antiapoptotic pathways, in the acute phase, and in the later stages reduce remodeling. Santos-Gallego et al. (2016) demonstrated how fingolimod induces significant Akt phosphorylation, ERK1/2, STAT3-β, and 24-hour GSK3 after a myocardial infarction, increased compared with controls. Therefore, in a porcine model of ischemia-reperfusion, fingolimod activated both RISK and SAFE in the acute phase of myocardial infarction.

The signal transduction pathways, including Akt and ERK1/2, are important stimuli for the growth and survival of cardiomyocytes. Although their short-term activation results in decreased apoptosis, in the long run, it leads to cardiomyocyte hypertrophy. Long-term activation of these pathways is one of the key molecular features to prevent post-infarction left ventricular remodeling (Haq et al., 2001; van Berlo, Maillet, & Molkentin, 2013). Evidence of the antiapoptotic action of fingolimod is further supported by detection of a reduced activation of proapoptotic proteins p53 and Caspase-3, the other to an antiapoptotic protein such as Bcl-2 expression increased and the kinase C-ε in pigs treated with fingolimod. As a result of antiapoptotic effects of fingolimod, this medication can significantly reduce the size of the infarcted area.

Attenuation of ischemic cardiomyopathy

Among other actions of fingolimod to attenuate myocardial damage, including reducing the extent of myocardial infarcted area acting on both the absolute and the percentage of the infarcted left ventricular myocardium. As evidenced by Santos-Gallego et al. (2016) fingolimod can increase the "rescue" infarction of approximately five times compared to controls that have not been given the drug.

Besides the reduction of the infarcted area, fingolimod has proven capable of improving left ventricular systolic function, as evidenced by increased left ventricular ejection fraction (LVEF), assessed by cardiac magnetic resonance, performed from a week to a month after myocardial infarction; also cases treated with fingolimod have greater contractile reserve compared with controls (Santos-Gallego et al., 2016). These findings are important because both improved LVEF and the presence of proper contractile reserve are associated with more favorable outcomes (Kelle et al., 2009; Williams, Odabashian, Lauer, Thomas, & Marwick, 1996).

One of the most negative complications of myocardial infarction is definitely the structural remodeling that follows, which is characterized by dilation, compensatory hypertrophy, and changes in left ventricular sphericity: Fingolimod reduces the extent of left ventricular remodeling, visible as reduced left ventricular mass 1 week to 1 month after myocardial infarction. In addition, compensatory hypertrophy of the wall, the wall thickness index calculated by echocardiography, is lower than in the group of pigs treated with fingolimod (Santos-Gallego et al., 2016).

Santos-Gallego et al. (2016) were the first to report that fingolimod not only reduces ischemia-reperfusion injury and the size of the infarcted area but also resizes the development of ischemic cardiomyopathy.

The fingolimod exert its remodelling effects by structural changes in left ventricular load, including reduced dilation of the heart echocardiography valuable both to the same room that the MRI. The antiremodeling is also visible at the histological level, as demonstrated by the reduced deposition of collagen in the myocardial interstitium fingolimod-treated group; this group also had small cardiomyocytes, confirming by microscope that fingolimod can alleviate postinfarct remodeling.

Role of sphingosine 1 phosphate-high-density lipoprotein in cardioprotection

About 60% of plasma is carried by high-density lipoprotein (HDL) (Morel et al., 2016). S1P, a component of HDL, induced effects that were similar to those of HDL. These compounds significantly reduced diffusion of fluorescent dye among cardiomyocytes ($\sim$50%), which could be prevented by PKC inhibition. As observed during optical recordings of transmembrane voltage, HDL and S1P depressed impulse conduction only minimally ($<$5%). Moreover, 5 minutes of HDL and S1P treatment at the onset of reperfusion significantly reduced infarct size ($\sim$50%) in response to 30 minutes ischemia in ex vivo experiments (Morel et al., 2016). More than 30 years ago it was shown that cholesterol carried by HDL is strongly and in inverse proportion linked to coronary artery disease. Recent studies assigning HDL properties that go far beyond capacity to promote only cholesterol efflux. They participate in lipoprotein transportation mediated by binding to S1P, transporter apoprotein M associated with HDL (apoM), and S1P is active in various HDL–associated signal transduction cascades through activation of its receptors on various cell types (James & Frias, 2014; Waeber & Walther, 2014).

HDL cardioprotective properties against coronary artery disease (CAD) have been attributed to the ability of S1P to preserve endothelial function and inhibit proinflammatory leukocyte level signaling (Poti, Simoni, & Nofer, 2014). This aspect is important because, in the evolution of myocardial infarction, accumulation of inflammatory cells in the infarcted area occurs rapidly shortly after initiation of the procedure of reperfusion and peak usually occurs 24 hours later neutrophils (Yan et al., 2013).

This book has identified a new potential pharmacological pre- and postconditioning drug for cardioprotection by attenuating ROS, inhibiting inflammatory pathways, and reducing apoptosis in myocardial ischemia-reperfusion injury.

References

Aguiar, C., Batista, S., & Pacheco, R. (2015). Cardiovascular effects of fingolimod: Relevance, detection and approach. *Revista Portuguesa de Cardiologia, 34*(4), 279–285. Available from https://doi.org/10.1016/j.repc.2014.11.012.

Ahmed, N., Linardi, D., Decimo, I., Mehboob, R., Gebrie, M. A., Innamorati, G., ... Rungatscher, A. (2017). Characterization and expression of sphingosine 1-phosphate receptors in human and rat heart. *Frontiers in Pharmacology, 8*, 312. Available from https://doi.org/10.3389/fphar.2017.00312.

Ahmed, N., Linardi, D., Muhammad, N., Chiamulera, C., Fumagalli, G., Biagio, L. S., ... Rungatscher, A. (2017). Sphingosine 1-phosphate receptor modulator fingolimod (FTY720) attenuates myocardial fibrosis in post-heterotopic heart transplantation. *Frontiers in Pharmacology, 8*, 645. Available from https://doi.org/10.3389/fphar.2017.00645.

Arish, M., Husein, A., Kashif, M., Saleem, M., Akhter, Y., & Rub, A. (2016). Sphingosine-1-phosphate signaling: unraveling its role as a drug target against infectious diseases. *Drug Discovery Today, 21*(1), 133–142. Available from https://doi.org/10.1016/j.drudis.2015.09.013.

Brinkmann, V. (2007). Sphingosine 1-phosphate receptors in health and disease: mechanistic insights from gene deletion studies and reverse pharmacology. *Pharmacology & Therapeutics, 115*(1), 84–105. Available from https://doi.org/10.1016/j.pharmthera.2007.04.006.

Brinkmann, V., Billich, A., Baumruker, T., Heining, P., Schmouder, R., Francis, G., ... Burtin, P. (2010). Fingolimod (FTY720): Discovery and development of an oral drug to treat multiple sclerosis. *Nature Reviews Drug Discovery, 9*(11), 883–897. Available from https://doi.org/10.1038/nrd3248.

Brinkmann, V., Davis, M. D., Heise, C. E., Albert, R., Cottens, S., Hof, R., ... Lynch, K. R. (2002). The immune modulator FTY720 targets sphingosine 1-phosphate receptors. *Journal of Biological Chemistry, 277*(24), 21453–21457. Available from https://doi.org/10.1074/jbc.C200176200.

Calabresi, P. A., Radue, E. W., Goodin, D., Jeffery, D., Rammohan, K. W., Reder, A. T., ... Lublin, F. D. (2014). Safety and efficacy of fingolimod in patients with relapsing-remitting multiple sclerosis (FREEDOMS II): A double-blind, randomised, placebo-controlled, phase 3 trial. *Lancet Neurology, 13*(6), 545–556. Available from https://doi.org/10.1016/s1474-4422(14)70049-3.

Chae, S. S., Proia, R. L., & Hla, T. (2004). Constitutive expression of the S1P1 receptor in adult tissues. *Prostaglandins & Other Lipid Mediators, 73*(1–2), 141–150.

Chiba, K. (2009). New therapeutic approach for autoimmune diseases by the sphingosine 1-phosphate receptor modulator, fingolimod (FTY720). *Yakugaku Zasshi, 129*(6), 655–665.

Chun, J., & Brinkmann, V. (2011). A mechanistically novel, first oral therapy for multiple sclerosis: The development of fingolimod (FTY720, Gilenya). *Discovery Medicine, 12* (64), 213–228.

Chun, J., & Hartung, H. P. (2010). Mechanism of action of oral fingolimod (FTY720) in multiple sclerosis. *Clinical Neuropharmacology, 33*(2), 91–101. Available from https://doi.org/10.1097/WNF.0b013e3181cbf825.

Cohen, J. A., & Chun, J. (2011). Mechanisms of fingolimod's efficacy and adverse effects in multiple sclerosis. *Annales of Neurology*, *69*(5), 759−777. Available from https://doi.org/10.1002/ana.22426.

Cohen, J. A., Barkhof, F., Comi, G., Hartung, H. P., Khatri, B. O., Montalban, X., ... Kappos, L. (2010). Oral fingolimod or intramuscular interferon for relapsing multiple sclerosis. *New England Journal of Medicine*, *362*(5), 402−415. Available from https://doi.org/10.1056/NEJMoa0907839.

Delbridge, M. S., Shrestha, B. M., Raftery, A. T., El Nahas, A. M., & Haylor, J. L. (2007). Reduction of ischemia-reperfusion injury in the rat kidney by FTY720, a synthetic derivative of sphingosine. *Transplantation*, *84*(2), 187−195. Available from https://doi.org/10.1097/01.tp.0000269794.74990.da.

Foster, C. A., Howard, L. M., Schweitzer, A., Persohn, E., Hiestand, P. C., Balatoni, B., ... Billich, A. (2007). Brain penetration of the oral immunomodulatory drug FTY720 and its phosphorylation in the central nervous system during experimental autoimmune encephalomyelitis: Consequences for mode of action in multiple sclerosis. *Journal of Pharmacology and Experimental Therapeutics*, *323*(2), 469−475. Available from https://doi.org/10.1124/jpet.107.127183.

Fukushima, N., Ishii, I., Contos, J. J., Weiner, J. A., & Chun, J. (2001). Lysophospholipid receptors. *Annual Review of Pharmacology and Toxicology*, *41*, 507−534. Available from https://doi.org/10.1146/annurev.pharmtox.41.1.507.

Gardell, S. E., Dubin, A. E., & Chun, J. (2006). Emerging medicinal roles for lysophospholipid signaling. *Trends in Molecular Medicine*, *12*(2), 65−75. Available from https://doi.org/10.1016/j.molmed.2005.12.001.

Gold, R., Comi, G., Palace, J., Siever, A., Gottschalk, R., Bijarnia, M., ... Kappos, L. (2014). Assessment of cardiac safety during fingolimod treatment initiation in a real-world relapsing multiple sclerosis population: A phase 3b, open-label study. *Journal of Neurology*, *261*(2), 267−276. Available from https://doi.org/10.1007/s00415-013-7115-8.

Gomez, L., Paillard, M., Price, M., Chen, Q., Teixeira, G., Spiegel, S., & Lesnefsky, E. J. (2011). A novel role for mitochondrial sphingosine-1-phosphate produced by sphingosine kinase-2 in PTP-mediated cell survival during cardioprotection. *Basic Research in Cardiology*, *106*(6), 1341−1353. Available from https://doi.org/10.1007/s00395-011-0223-7.

Graler, M. H., & Goetzl, E. J. (2004). The immunosuppressant FTY720 down-regulates sphingosine 1-phosphate G-protein-coupled receptors. *FASEB Journal*, *18*(3), 551−553. Available from https://doi.org/10.1096/fj.03-0910fje.

Groves, A., Kihara, Y., & Chun, J. (2013). Fingolimod: Direct CNS effects of sphingosine 1-phosphate (S1P) receptor modulation and implications in multiple sclerosis therapy. *Journal of the Neurological Sciences*, *328*(1−2), 9−18. Available from https://doi.org/10.1016/j.jns.2013.02.011.

Haq, S., Choukroun, G., Lim, H., Tymitz, K. M., del Monte, F., Gwathmey, J., ... Molkentin, J. D. (2001). Differential activation of signal transduction pathways in human hearts with hypertrophy versus advanced heart failure. *Circulation*, *103*(5), 670−677.

Heusch, G. (2013). Cardioprotection: Chances and challenges of its translation to the clinic. *Lancet*, *381*(9861), 166−175. Available from https://doi.org/10.1016/s0140-6736(12)60916-7.

Heusch, G. (2015). Molecular basis of cardioprotection: Signal transduction in ischemic pre-, post-, and remote conditioning. *Circulation Research*, *116*(4), 674−699. Available from https://doi.org/10.1161/circresaha.116.305348.

Heusch, G., Musiolik, J., Gedik, N., & Skyschally, A. (2011). Mitochondrial STAT3 activation and cardioprotection by ischemic postconditioning in pigs with regional myocardial ischemia/reperfusion. *Circulation Research*, *109*(11), 1302−1308. Available from https://doi.org/10.1161/circresaha.111.255604.

Hofmann, U., Burkard, N., Vogt, C., Thoma, A., Frantz, S., Ertl, G., ... Bonz, A. (2009). Protective effects of sphingosine-1-phosphate receptor agonist treatment after myocardial ischaemia-reperfusion. *Cardiovascular Research*, *83*(2), 285−293. Available from https://doi.org/10.1093/cvr/cvp137.

Hsu, Y. H., Hirota, A., Shima, S., Nakagawa, M., Adachi, T., Nozaki, H., & Nakayama, M. (1989). Myrocin C, a new diterpene antitumor antibiotic from *Myrothecium verrucaria*. II. Physico-chemical properties and structure determination. *Journal of Antibiotics (Tokyo)*, *42*(2), 223−229.

James, R. W., & Frias, M. A. (2014). High density lipoproteins and ischemia reperfusion injury: The therapeutic potential of HDL to modulate cell survival pathways. *Advances in Experimental Medicine and Biology*, *824*, 19−26. Available from https://doi.org/10.1007/978-3-319-07320-0_3.

Jin, Z. Q., Zhang, J., Huang, Y., Hoover, H. E., Vessey, D. A., & Karliner, J. S. (2007). A sphingosine kinase 1 mutation sensitizes the myocardium to ischemia/reperfusion injury. *Cardiovascular Research*, *76*(1), 41−50. Available from https://doi.org/10.1016/j.cardiores.2007.05.029.

Jin, Z. Q., Zhou, H. Z., Zhu, P., Honbo, N., Mochly-Rosen, D., Messing, R. O., ... Gray, M. O. (2002). Cardioprotection mediated by sphingosine-1-phosphate and ganglioside GM-1 in wild-type and PKC epsilon knockout mouse hearts. *American Journal of Physiology-Heart and Circulatory Physiology*, *282*(6), H1970−H1977. Available from https://doi.org/10.1152/ajpheart.01029.2001.

Kappos, L., Radue, E. W., O'Connor, P., Polman, C., Hohlfeld, R., Calabresi, P., ... Burtin, P. (2010). A placebo-controlled trial of oral fingolimod in relapsing multiple sclerosis. *New England Journal of Medicine*, *362*(5), 387−401. Available from https://doi.org/10.1056/NEJMoa0909494.

Karliner, J. S. (2013). Sphingosine kinase and sphingosine 1-phosphate in the heart: A decade of progress. *Biochimica et Biophysica Acta*, *1831*(1), 203−212. Available from https://doi.org/10.1016/j.bbalip.2012.06.006.

Karliner, J. S., Honbo, N., Summers, K., Gray, M. O., & Goetzl, E. J. (2001). The lysophospholipids sphingosine-1-phosphate and lysophosphatidic acid enhance survival during hypoxia in neonatal rat cardiac myocytes. *Journal of Molecular and Cellular Cardiology*, *33*(9), 1713−1717. Available from https://doi.org/10.1006/jmcc.2001.1429.

Kelle, S., Roes, S. D., Klein, C., Kokocinski, T., de Roos, A., Fleck, E., ... Nagel, E. (2009). Prognostic value of myocardial infarct size and contractile reserve using magnetic resonance imaging. *Journal of the American College of Cardiology*, *54*(19), 1770−1777. Available from https://doi.org/10.1016/j.jacc.2009.07.027.

Kelly, R. F., Lamont, K. T., Somers, S., Hacking, D., Lacerda, L., Thomas, P., ... Lecour, S. (2010). Ethanolamine is a novel STAT-3 dependent cardioprotective agent.

Basic Research in Cardiology, *105*(6), 763–770. Available from https://doi.org/10.1007/s00395-010-0125-0.

Kennedy, S., Kane, K. A., Pyne, N. J., & Pyne, S. (2009). Targeting sphingosine-1-phosphate signalling for cardioprotection. *Current Opinion in Pharmacology*, *9*(2), 194–201. Available from https://doi.org/10.1016/j.coph.2008.11.002.

Khatri, B., Barkhof, F., Comi, G., Hartung, H. P., Kappos, L., Montalban, X., ... Cohen, J. A. (2011). Comparison of fingolimod with interferon beta-1a in relapsing-remitting multiple sclerosis: A randomised extension of the TRANSFORMS study. *Lancet Neurology*, *10*(6), 520–529. Available from https://doi.org/10.1016/s1474-4422(11)70099-0.

Kovarik, J. M., Riviere, G. J., Neddermann, D., Maton, S., Hunt, T. L., & Schmouder, R. L. (2008). A mechanistic study to assess whether isoproterenol can reverse the negative chronotropic effect of fingolimod. *Journal of Clinical Pharmacology*, *48*(3), 303–310. Available from https://doi.org/10.1177/0091270007312903.

Kovarik, J. M., Slade, A., Riviere, G. J., Neddermann, D., Maton, S., Hunt, T. L., & Schmouder, R. L. (2008). The ability of atropine to prevent and reverse the negative chronotropic effect of fingolimod in healthy subjects. *British Journal of Clinical Pharmacology*, *66*(2), 199–206. Available from https://doi.org/10.1111/j.1365-2125.2008.03199.x.

Koyrakh, L., Roman, M. I., Brinkmann, V., & Wickman, K. (2005). The heart rate decrease caused by acute FTY720 administration is mediated by the G protein-gated potassium channel I. *American Journal of Transplantation*, *5*(3), 529–536. Available from https://doi.org/10.1111/j.1600-6143.2005.00754.x.

Lebrun, C., Vermersch, P., Brassat, D., Defer, G., Rumbach, L., Clavelou, P., ... Frenay, M. (2011). Cancer and multiple sclerosis in the era of disease-modifying treatments. *Journal of Neurology*, *258*(7), 1304–1311. Available from https://doi.org/10.1007/s00415-011-5929-9.

Lecour, S., Smith, R. M., Woodward, B., Opie, L. H., Rochette, L., & Sack, M. N. (2002). Identification of a novel role for sphingolipid signaling in TNF alpha and ischemic preconditioning mediated cardioprotection. *Journal of Molecular and Cellular Cardiology*, *34*(5), 509–518. Available from https://doi.org/10.1006/jmcc.2002.1533.

Liu, Y., Wada, R., Yamashita, T., Mi, Y., Deng, C. X., Hobson, J. P., ... Proia, R. L. (2000). Edg-1, the G protein-coupled receptor for sphingosine-1-phosphate, is essential for vascular maturation. *Journal of Clinical Investigation*, *106*(8), 951–961. Available from https://doi.org/10.1172/jci10905.

Man, K., Ng, K. T., Lee, T. K., Lo, C. M., Sun, C. K., Li, X. L., ... Fan, S. T. (2005). FTY720 attenuates hepatic ischemia-reperfusion injury in normal and cirrhotic livers. *American Journal of Transplantation*, *5*(1), 40–49. Available from https://doi.org/10.1111/j.1600-6143.2004.00642.x.

Means, C. K., & Brown, J. H. (2009). Sphingosine-1-phosphate receptor signalling in the heart. *Cardiovascular Research*, *82*(2), 193–200. Available from https://doi.org/10.1093/cvr/cvp086.

Means, C. K., Xiao, C. Y., Li, Z., Zhang, T., Omens, J. H., Ishii, I., ... Brown, J. H. (2007). Sphingosine 1-phosphate S1P2 and S1P3 receptor-mediated Akt activation protects against in vivo myocardial ischemia-reperfusion injury. *American Journal of Physiology-Heart and Circulatory Physiology*, *292*(6), H2944–H2951. Available from https://doi.org/10.1152/ajpheart.01331.2006.

Morel, S., Christoffersen, C., Axelsen, L. N., Montecucco, F., Rochemont, V., Frias, M. A., ... Kwak, B. R. (2016). Sphingosine-1-phosphate reduces ischaemia-reperfusion injury by phosphorylating the gap junction protein Connexin43. *Cardiovascular Research*, *109*(3), 385−396. Available from https://doi.org/10.1093/cvr/cvw004.

Noguchi, K., & Chun, J. (2011). Roles for lysophospholipid S1P receptors in multiple sclerosis. *Critical Reviews in Biochemistry and Molecular Biology*, *46*(1), 2−10. Available from https://doi.org/10.3109/10409238.2010.522975.

Oo, M. L., Thangada, S., Wu, M. T., Liu, C. H., Macdonald, T. L., Lynch, K. R., ... Hla, T. (2007). Immunosuppressive and anti-angiogenic sphingosine 1-phosphate receptor-1 agonists induce ubiquitinylation and proteasomal degradation of the receptor. *Journal of Biological Chemistry*, *282*(12), 9082−9089. Available from https://doi.org/10.1074/jbc.M610318200.

Peters, S. L., & Alewijnse, A. E. (2007). Sphingosine-1-phosphate signaling in the cardiovascular system. *Current Opinion in Pharmacology*, *7*(2), 186−192. Available from https://doi.org/10.1016/j.coph.2006.09.008.

Piot, C., Croisille, P., Staat, P., Thibault, H., Rioufol, G., Mewton, N., ... Ovize, M. (2008). Effect of cyclosporine on reperfusion injury in acute myocardial infarction. *New England Journal of Medicine*, *359*(5), 473−481. Available from https://doi.org/10.1056/NEJMoa071142.

Poti, F., Simoni, M., & Nofer, J. R. (2014). Atheroprotective role of high-density lipoprotein (HDL)-associated sphingosine-1-phosphate (S1P). *Cardiovascular Research*, *103*(3), 395−404. Available from https://doi.org/10.1093/cvr/cvu136.

Rosen, H., & Goetzl, E. J. (2005). Sphingosine 1-phosphate and its receptors: An autocrine and paracrine network. *Nature Reviews Immunology*, *5*(7), 560−570. Available from https://doi.org/10.1038/nri1650.

Salomone, S., Yoshimura, S., Reuter, U., Foley, M., Thomas, S. S., Moskowitz, M. A., & Waeber, C. (2003). S1P3 receptors mediate the potent constriction of cerebral arteries by sphingosine-1-phosphate. *European Journal of Pharmacology*, *469*(1−3), 125−134.

Santos-Gallego, C. G., Vahl, T. P., Goliasch, G., Picatoste, B., Arias, T., Ishikawa, K., ... Badimon, J. J. (2016). Sphingosine-1-phosphate receptor agonist fingolimod increases myocardial salvage and decreases adverse postinfarction left ventricular remodeling in a porcine model of ischemia/reperfusion. *Circulation*, *133*(10), 954−966. Available from https://doi.org/10.1161/circulationaha.115.012427.

Schmouder, R., Serra, D., Wang, Y., Kovarik, J. M., DiMarco, J., Hunt, T. L., & Bastien, M. C. (2006). FTY720: Placebo-controlled study of the effect on cardiac rate and rhythm in healthy subjects. *Journal of Clinical Pharmacology*, *46*(8), 895−904. Available from https://doi.org/10.1177/0091270006289853.

Schwab, S. R., & Cyster, J. G. (2007). Finding a way out: Lymphocyte egress from lymphoid organs. *Nature Immunology*, *8*(12), 1295−1301. Available from https://doi.org/10.1038/ni1545.

Singer, I. I., Tian, M., Wickham, L. A., Lin, J., Matheravidathu, S. S., Forrest, M. J., ... Quackenbush, E. J. (2005). Sphingosine-1-phosphate agonists increase macrophage homing, lymphocyte contacts, and endothelial junctional complex formation in murine lymph nodes. *Journal of Immunology*, *175*(11), 7151−7161.

Skyschally, A., Schulz, R., & Heusch, G. (2010). Cyclosporine A at reperfusion reduces infarct size in pigs. *Cardiovascular Drugs and Therapy*, *24*(1), 85−87. Available from https://doi.org/10.1007/s10557-010-6219-y.

Spiegel, S., & Milstien, S. (2003). Sphingosine-1-phosphate: An enigmatic signalling lipid. *Nature Reviews Molecular Cell Biology*, *4*(5), 397−407. Available from https://doi.org/10.1038/nrm1103.

Tao, R., Zhang, J., Vessey, D. A., Honbo, N., & Karliner, J. S. (2007). Deletion of the sphingosine kinase-1 gene influences cell fate during hypoxia and glucose deprivation in adult mouse cardiomyocytes. *Cardiovascular Research*, *74*(1), 56−63. Available from https://doi.org/10.1016/j.cardiores.2007.01.015.

Theilmeier, G., Schmidt, C., Herrmann, J., Keul, P., Schafers, M., Herrgott, I., ... Levkau, B. (2006). High-density lipoproteins and their constituent, sphingosine-1-phosphate, directly protect the heart against ischemia/reperfusion injury in vivo via the S1P3 lysophospholipid receptor. *Circulation*, *114*(13), 1403−1409. Available from https://doi.org/10.1161/circulationaha.105.607135.

Tolle, M., Levkau, B., Keul, P., Brinkmann, V., Giebing, G., Schonfelder, G., ... Van., & der Giet, M. (2005). Immunomodulator FTY720 Induces eNOS-dependent arterial vasodilatation via the lysophospholipid receptor S1P3. *Circulation Research*, *96*(8), 913−920. Available from https://doi.org/10.1161/01.res.0000164321.91452.00.

van Berlo, J. H., Maillet, M., & Molkentin, J. D. (2013). Signaling effectors underlying pathologic growth and remodeling of the heart. *Journal of Clinical Investigation*, *123*(1), 37−45. Available from https://doi.org/10.1172/jci62839.

Wacker, B. K., Park, T. S., & Gidday, J. M. (2009). Hypoxic preconditioning-induced cerebral ischemic tolerance: Role of microvascular sphingosine kinase 2. *Stroke*, *40*(10), 3342−3348. Available from https://doi.org/10.1161/strokeaha.109.560714.

Waeber, C., & Walther, T. (2014). Sphingosine-1-phosphate as a potential target for the treatment of myocardial infarction. *Circulation Journal*, *78*(4), 795−802.

Waeber, C., Blondeau, N., & Salomone, S. (2004). Vascular sphingosine-1-phosphate S1P1 and S1P3 receptors. *Drug News & Perspectives*, *17*(6), 365−382.

Walsh, K. B. (2011). Targeting GIRK channels for the development of new therapeutic agents. *Frontiers in Pharmacology*, *2*, 64. Available from https://doi.org/10.3389/fphar.2011.00064.

Wang, G., Kim, R. Y., Imhof, I., Honbo, N., Luk, F. S., Li, K., ... Raffai, R. L. (2014). The immunosuppressant FTY720 prolongs survival in a mouse model of diet-induced coronary atherosclerosis and myocardial infarction. *Journal of Cardiovascular Pharmacology and Therapeutics*, *63*(2), 132−143. Available from https://doi.org/10.1097/fjc.0000000000000031.

Wei, Y., Yemisci, M., Kim, H. H., Yung, L. M., Shin, H. K., Hwang, S. K., ... Waeber, C. (2011). Fingolimod provides long-term protection in rodent models of cerebral ischemia. *Annals of Neurology*, *69*(1), 119−129. Available from https://doi.org/10.1002/ana.22186.

Williams, M. J., Odabashian, J., Lauer, M. S., Thomas, J. D., & Marwick, T. H. (1996). Prognostic value of dobutamine echocardiography in patients with left ventricular dysfunction. *Journal of the American College of Cardiology*, *27*(1), 132−139. Available from https://doi.org/10.1016/0735-1097(95)00393-2.

Xiang, S. Y., Ouyang, K., Yung, B. S., Miyamoto, S., Smrcka, A. V., Chen, J., & Heller Brown, J. (2013). PLCepsilon, PKD1, and SSH1L transduce RhoA signaling to protect mitochondria from oxidative stress in the heart. *Science Signal*, *6*(306), ra108. Available from https://doi.org/10.1126/scisignal.2004405.

Yan, X., Anzai, A., Katsumata, Y., Matsuhashi, T., Ito, K., Endo, J., . . . Sano, M. (2013). Temporal dynamics of cardiac immune cell accumulation following acute myocardial infarction. *Journal of Molecular and Cellular Cardiology, 62*, 24−35. Available from https://doi.org/10.1016/j.yjmcc.2013.04.023.

Yang, Z., Day, Y. J., Toufektsian, M. C., Xu, Y., Ramos, S. I., Marshall, M. A., . . . Linden, J. (2006). Myocardial infarct-sparing effect of adenosine A2A receptor activation is due to its action on CD4 + T lymphocytes. *Circulation, 114*(19), 2056−2064. Available from https://doi.org/10.1161/circulationaha.106.649244.

Yellon, D. M., & Hausenloy, D. J. (2007). Myocardial reperfusion injury. *New England Journal of Medicine, 357*(11), 1121−1135. Available from https://doi.org/10.1056/NEJMra071667.

Zhang, J., Honbo, N., Goetzl, E. J., Chatterjee, K., Karliner, J. S., & Gray, M. O. (2007). Signals from type 1 sphingosine 1-phosphate receptors enhance adult mouse cardiac myocyte survival during hypoxia. *American Journal of Physiology-Heart and Circulatory Physiology, 293*(5), H3150−H3158. Available from https://doi.org/10.1152/ajpheart.00587.2006.

Index

Note: Page numbers followed by "*f*" refer to figures.